Introduction to Naval Architecture

Introduction to Naval Architecture

Editor

Alfred Eckstein

scitus academics

Introduction to Naval Architecture
Edited by **Alfred Eckstein**

ISBN: 978-1-68117-457-0
Library of Congress Control Number: 2016937627

Notice

Printed in United States of America on Acid Free Paper

Preface

For many people involved in the design, construction, operation and maintenance of all marine vessels, the basic characteristics of a ship's design and how they affect its behavior at sea are of crucial importance. Naval architects and ship designers must have a thorough understanding of these principles. Marine engineers also need to recognize the extent to which these principles influence and limit their activities. Naval architecture is an engineering discipline that deals with the design, construction, maintenance and operation of ships and structures. Naval architecture involves basic and applied research, design, development, evaluation of design and calculation at all stages of a marine vehicle life. The main activities involved are preliminary design, detailed design, construction, testing, operation and maintenance, launch and dry docking. Knowledge of the basics of naval architecture is essential not only for newcomers on the ground, but also for the wealth of marine architects, including marine engineers, marine surveyors and ship crews. This book presents wide-ranging coverage to the subject, offering a clear and concise understanding of the basics of this broad field. This book is ideal for those who first approach the subject and those who want to update or refresh their knowledge in areas further than their direct expertise.

Table of Contents

CHAPTER 1

Naval ship's susceptibility assessment by the probabilistic density function

Kwang Sik Kim, Se Yun Hwang and Jang Hyun Lee*

Department of Naval Architecture and Ocean Engineering, INHA University, 100 Inharo, Nam-gu, Incheon 402-751, Republic of Korea

ABSTRACT

The survivability of the naval ship is the capability of a warship to avoid or withstand a hostile environment. The survivability of the naval ship assessed by three categories (susceptibility, vulnerability and recoverability). The magnitude of susceptibility of a warship encountering with threat is dependent upon the attributes of detection equipment and weapon system. In this paper, as a part of a naval ship's survivability analysis, an assessment process model for the ship's susceptibility analysis technique is developed. Naval ship's survivability emphasizing the susceptibility is assessed by the probability of detection, and the probability of hit. Considering the radar cross section (RCS), the assessment procedure for the susceptibility is described. It's emphasizing the simplified calculation model based on the probability density function for probability of hit. Assuming the probability of hit given a both single-hit and multiple-hit, the susceptibility is accessed for a RCS and the hit probability for a rectangular target is applied for a given threat.

KEYWORDS

Survivability; Vulnerability; RCS (Radar Cross Section); Probability of hit; Probability of detection

1. INTRODUCTION

When a naval ship in a modern combat environment is exposed to a threatening weapon and attacked, the combat system and hull structure suffer critical damage. The ability of a naval ship to withstand such threats encountered in a battle environment is defined as the survivability of the ship [1]. The survivability of the ship is evaluated by three types of stochastic indicators—susceptibility, vulnerability, and recoverability [1], [2] and [3]. The most fundamental method for improving the survivability of the naval ship is to design the ship such that its susceptibility becomes close to zero. However, because it is difficult to attain zero susceptibility in reality, the realistic approach to improving the susceptibility involves considering various situations that may arise on being attacked [4]. The susceptibility of the naval ship refers to the probability of the ship being attacked by threatening weapons after being identified by an enemy's detection technology and equipment [4] and [5]. Therefore, the signals generated by a naval ship, the devices for detecting and identifying these signals, and the defense and deception system employed against enemy's threatening weapons following detection need to be analyzed [6]. The susceptibility should be comprehensively evaluated by considering the shape of the naval ship's radar reflection area, infrared signal characteristics, electromagnetic equipment performance, and radiation noises [7]. Many studies have reported on the aforementioned evaluation factors indivi-dually [8], and the analysis field can be divided into Radar Cross Section (RCS), Infrared (IR), Underwater Radiated Noise (URN), and Electro Magnetic Interference (EMI) according to the respective detection characteristics and equipment [6]. In the present study, as a part of a naval ship's survivability analysis, a theoretical process model for the ship's susceptibility analysis technique was developed. The applicability of this model was demonstrated through an example model and arbitrary attack situations.

2. SUSCEPTIBILITY ANALYSIS OF NAVAL SHIPS

2.1. Susceptibility analysis theory

To analyze the susceptibility of a naval ship, the characteristics and performance of the equipment used for detecting enemy naval ships, and the possibility of being hit by threatening weapons must be analyzed. In this study, the susceptibility analysis process model was constituted as shown in Figure 1 based on the stochastic theory. The attacking probability (P_H) is based on the detection probability (P_D), which is the probability of identifying the friendly naval ship by the enemy detection equipment, and on the hit probability (P_{hit}), which is the probability that the friendly naval ship could be hit by the threatening weapon after being detected. The attacking probability is expressed as shown in Eq. (1).

$$P_H = P_D \cdot P_{hit} \tag{1}$$

2.1.1. Detection probability

The detection probability is the probability that the naval ship (target) could be detected by the enemy detection equipment. The RCS detection characteristics and the threshold to noise (T/N) value, which is the minimum detection limit, are calculated on the basis of the signal to noise ratio (S/N) value. The S/N value is the degree of noise in RCS signal according to the cross sectional area (σ) of the radar [9] (Eq. (2)). RCS is the specified area to indicate the reflectivity of the radar's reflector when the electromagnetic waves emitted from the radar are reflected back by the target [7]. Therefore, the reflectivity value of the radar varies depending on the shape of the reflector [7]. Naval systems, including the naval ship's hull and fittings, have their own unique complex shapes, which have to be taken into account when calculating RCS; hence, each shape should be simplified as shown in Figure 2, and then, RCS should be calculated for each simplified shape [9].

$$P_D = \left[1 + \frac{2(T/N)(S/N)}{2 + (S/N)^2}\right] e^{-2(T/N)(2+S/N)} \tag{2}$$

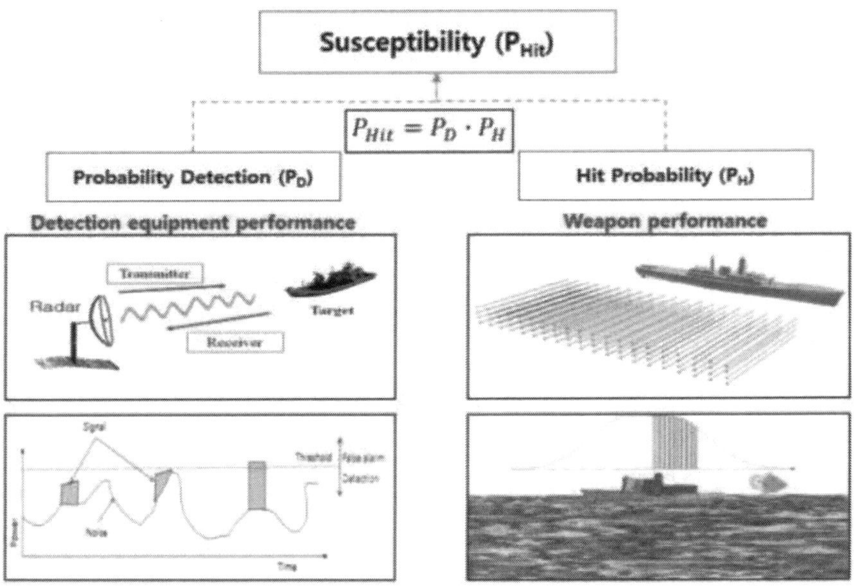

Figure 1. Mesh grid of topographic model.

S/N ratio value (dB) represents the extent of unnecessary noise in the signal [10] and [12]. In other words, the S/N ratio represents radar signal-to-noise ratio (Figure 3), and it is proportional to the shape of RCS because it is calculated using the radar's cross-sectional area. The S/N value according to the RCS detection characteristics is defined as shown in Eq. (3); the variables are defined in Table 1.

$$S/N = \left[\frac{PG^2\lambda^2\sigma}{(4\pi)^3 R^4 FkTB_n L}\right] \tag{3}$$

Here, radar peak power indicates the antenna reception power, and kTB_n value indicates noise power (Eq. (4)) [10].

$$N(Noise\ power) = kTB_n \tag{4}$$

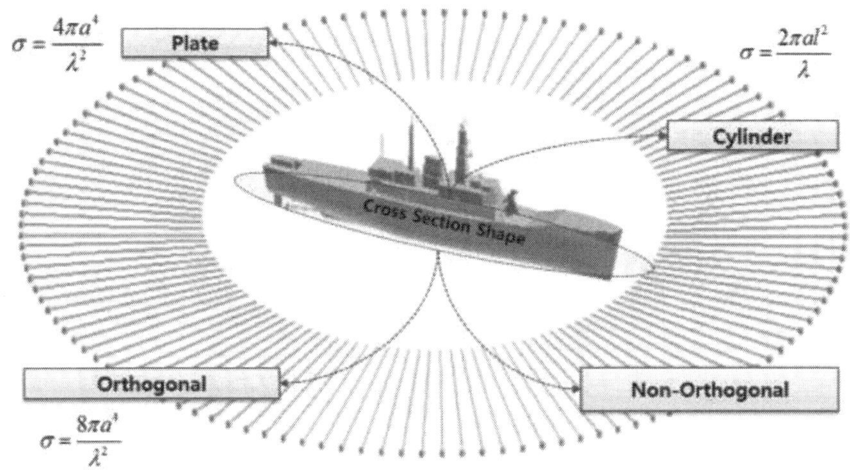

Figure 2. The simplified shape for hull and system.

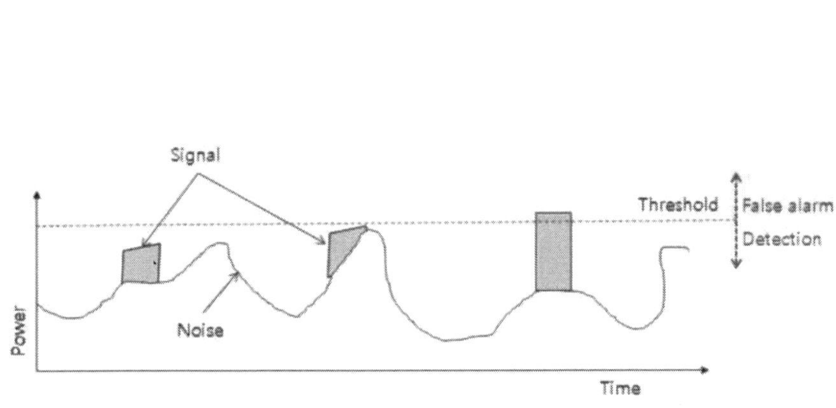

Figure 3. Conditional detection by threshold setting.

As shown in Figure 2, the threshold to noise ratio (T/N) indicates the minimum signal threshold for detection [11], and it is calculated using Eq. (5). In Eq. (5), P_{false} indicates the probability of error (false alarm) when the enemy detection equipment identifies a target. In fact, it is difficult to calculate the error probability (P_{false}) accurately because these data are obtained through Live Fire Testing and Evaluation (LFT&E). In the pre-sent study, the detection probability is calculated by specifying arbitrary error probability.

$$N(Noise\ power) = kTB_n \tag{5}$$

Table 1. Definition of variable for Eq. (3).

Variable	Definition	Dimension
P	Radar peak power	[Watt]
R	Distance from radar to target	[m]
λ	Wavelength of signal	[m]
G	Antenna gain factor	Const.
K	Boltmann's constant	1.381 × 10-23 [J/deg]
N	Noise factor	Const.
T	Temperature	[°C]
Bn	Radar Receiver bandwidth	[Hz]
L	Signal echo power loss factor	Const.

2.1.2. Hit probability

Hit probability is the probability that the enemy threat weapon, which is targeted at the friendly ship, hits the target area, and it is calculated according to the characteristics and effectiveness of the threat weapon [5].

In the present study, the hit probability was calculated for two-dimensional target area using the probability density function under the assumption the threat weapon characteristics are classified as single hit and multiple hit according to the number of hits by the war head, as mentioned in Figure 1. The procedure to calculate the hit probability for a single hit is as follows:

(1) First, the target area (T_a) for the probability analysis is defined; this target area is an arbitrary area (T_a) from the total area of the naval ship.

(2) The probability distribution of the hit locations in the length and depth direction of the naval ship is defined. The data on probability distribution of the hit locations are obtained through the actual LFT&E. In the case of the naval ship survivability analysis program, MOTISS™, the probability distributions of the hit location in the length and depth direction of the naval ship are defined by the normal distribution and Weibull distribution, respectively [6]. In the present study, the hit probability was calculated by assuming that the probability distributions of each hit location in the length and depth directions of the naval ship as normal distributions (Figure 4).

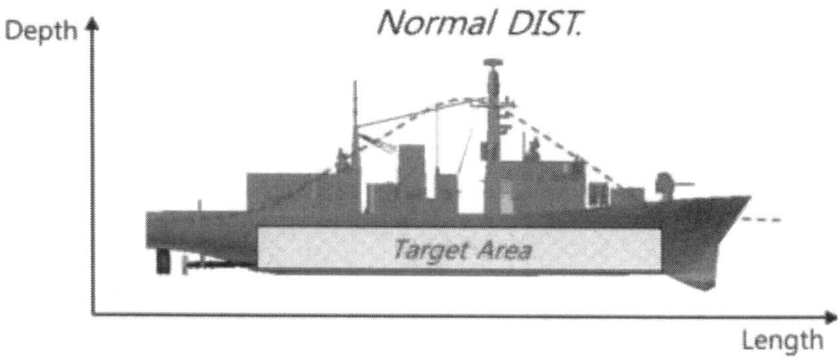

Figure 4. The example of probability distribution for hit on target area.

(3) Thereafter, the area of the probability density function (PDF) with regard to the target area of analysis is calculated by the hit probability (Figure 5). The hit probability of a single hit can be defined by Eqs. (6), (7), (8) and (9). In these equations, x and y represent the variables in the length and depth direction, respectively, x_1~ x_2 and y_1~ y_2 indicate the hit area section in each plane. The linear error probability (LEP) of identifying the threat weapon's target in the normal-distribution-based hit shape of the naval ship is defined by Eq. (11). In the calculation of the hit probability with LEP, μ and σ represent the mean and standard deviations, respectively, of the corresponding probability distributions.

$$P_H(x,y) = P(Langth) \cdot P(Depth)$$
$$= PDF(x) \cdot PDF(y) \tag{6}$$

$$PDF(x) \cdot PDF(y) = \int_{-\frac{a}{2}}^{\frac{a}{2}} P(x)dx \cdot \int_{-\frac{b}{2}}^{\frac{b}{2}} P(y)dy \tag{7}$$

$$P(x_1 < x < x_2) = \int_{x_1}^{x_2} \frac{1}{\sqrt{2\pi}\sigma} exp\left[-\frac{(x-\mu)^2}{\sigma^2}\right] dx \tag{8}$$

$$P(y_1 < y < y_2) = \int_{y_1}^{y_2} \frac{1}{\sqrt{2\pi}\sigma} exp\left[-\frac{(y-\mu)^2}{\sigma^2}\right] dy \tag{9}$$

$$LEP = 0.6745\sigma \tag{10}$$

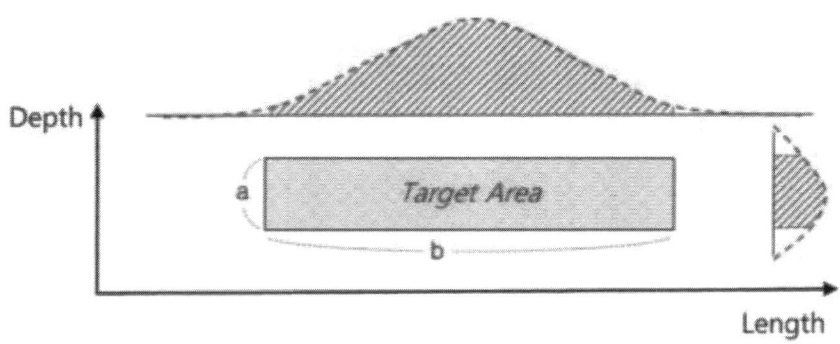

Figure 5. The probability distribution for target area on 2-dimensional surface.

The hit probability of multiple hits is determined by the total number of hits as calculated by using the hit probability calculation equation for a single hit, which was defined earlier [5]. The procedure for calculating the hit probability of multiple hits is as follows. (1) First, the total number of hits on the entire naval ship is determined (Eq. (11)). (2) The expected value ($E(x)$) of the hits on the pertinent target area is calculated (Eq. (12)). (3) Finally, the hit probability is calculated by Eq. (13) according to the expected hit values.

$$M = \frac{A_p}{A_c} \cdot K \quad (A_c = range^2\theta) \tag{11}$$

$$E(x) = \frac{1}{\sqrt{2\pi\sigma^2}} exp\left[-\frac{(x-\mu)^2}{2\sigma^2}\right] \tag{12}$$

$$P_H(H \geq 1) = \sum_{n=1}^{n} \left(1 - P_{single}\right)^{n-1} \cdot P_{single}$$
$$= P_{single} \frac{1-\left(1-P_{single}\right)^n}{1-\left(1-P_{single}\right)} = 1-\left(1-P_{single}\right)^n \tag{13}$$

2.2. Example of susceptibility analysis

In this section, an example model of the naval ship susceptibility analysis is described according to arbitrary multiple penetration threats and RCS detection characteristics. For this example model, the susceptibility analysis of arbitrary target areas, including the propulsion system, was performed by assuming that the radar detection is expected to have the attributes listed in Table 2 for a naval ship whose overall length and the height of the hull outer plate exposed above the water surface are 120 m and 2 m, respectively. Furthermore, to calculate the detection probability in the susceptibility analysis, the naval ships' outer plate was assumed to be a flat plate, and the susceptibility analysis was performed for the hull outer plate excluding the super structure on the main deck.

Table 2. Detection properties for radar.

Property	Value	Dimension
Receiver sensitivity	1E-14	W(m)
Operation frequency	0.25	[GHz]
Distance from radar	25,000	[m]
Angle	90	[θ]
Noise factor	2.5E-16	[w]
Temperature	195	[F]

Eq. (14) shows the calculation of the RCS value (σ) of the flat plate shape that was assumed as an example in this study. Using this value, the S/N value depending on the RCS detection is calculated, as shown in Eq. (15). Next, the detection probability is calculated based on Eq. (3) by considering the T/N and S/N values (Eq. (16)), where the T/N value is calculated using Eq. (17) by assuming an error probability (T_{false}) of 10^{-3}.

$$\sigma = \frac{4\pi(240)^2}{1.2^2} = 502{,}400$$
$$(With\ wave\ length = 1.2) \tag{14}$$

$$S/N = \frac{PG^2\lambda^2\sigma}{(4\pi)^3 R^4 FkTB_n L} = 12.81 \tag{15}$$

$$P_D = \left[1 + \frac{2(6.91)(12.81)}{2 + (12.81)^2}\right] e^{-2(6.91)(2+12.81)}$$
$$= 0.813 \tag{16}$$

$$\frac{T}{N} = -\ln(0.001) = 6.91 \tag{17}$$

In the present example model, the threat weapon whose attributes are as shown in Figure 6 was assumed to be the fragments from a near-explosion hitting the naval ship. Eq. (18) is the calculation example for the number of hits (*M*) on the pertinent target areas, including the propulsion system. The distance range, incidence angle (*θ*), and total number of hits (*K*) were as-sumed to be 80 m, 90°, and 70, respectively.

$$\mathrm{M} = \frac{500}{10048} \times 70 = 3.5 \tag{18}$$

With regard to the number of hits calculated by Eq. (18), the hit probability, which is based on the PDF, is calculated by Eq. (19). Therefore, the final hit probability of the present example model is calculated as shown in Eq. (20) by using the detection probability (T_D) and hit probability (T_{hit}) calculated earlier. In other words, the final hit probability is identified as approximately 70% during multiple hits on a 120 m long naval ship, i.e., the present example model.

$$P_H(H \geq 3.48) = 1 - (1 - 0.446)^{3.48} = 0.872 \quad (19)$$

$$P_H = 0.813 \times 0.872 = 0.709 \quad (20)$$

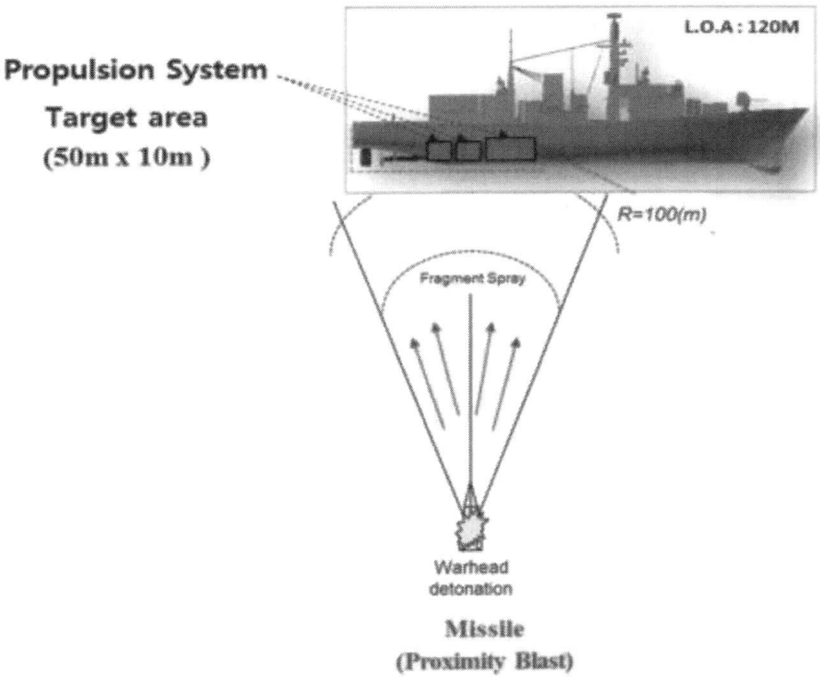

Figure 6. The example of threat scenario.

3. CONCLUSIONS

In the present study, a theoretical procedure was proposed to analyze quantitatively the susceptibility, which is one of the analysis components of the naval ship's survivability. The susceptibility analysis procedure was divided into two steps determining the detection probability that is based on the characteristics of the detection equipment, and determining the hit probability that is based on the characteristics of the threat weapon. In particular, the hit probability calculation procedure for single hit and multiple hits by the threat weapon was proposed by using the two-dimensional PDF. Furthermore, the susceptibility analysis procedures were organized for RCS detection and dealing with multiple penetration using examples based on the proposed theory. However, the detection characteristics, which are identified upon exposure of the naval ship in a battle environment, vary depending on the detection equipment; hence, the detection probability calculation must be formulated based on the S/N value for various detection characteristics such as infra-red and underwater radiated noise. In addition, the probability distribution, which is the core of the present susceptibility analysis procedure, is assumed as a normal distribution; in future, a probability distribution model with a quantitative base should be established.

ACKNOWLEDGEMENTS

The present study is a part of the project "Analysis technique study (ST-54) for improving integrated survivabili-ty of naval ship – the Specialized Research Center for Survivability Technology," which is supported by the Defense Acquisition Program Administration and the National Defense and Science Institute (Contract number: UD120019OD) and supported by a Special Education Program for Offshore Plant by the Ministry of Trade, Industry and Energy Affairs (MOTIE).

REFERENCES

1. Ball RE. The fundamentals of aircraft combat survivability analysis and design. 2nd ed. Reston(Virginia): AIAA Education Series; 1985. 894 p.

2. Kim KH, Kim JH, Choi TM, Kim YH, Cho DS. A study on a dynamic radar cross section analysis technique for a surface warship. Journal of Ocean Engineering and Technology. 2009; 23(6): 77-81.

3. Ch YJ. A study on measuring procedure and analysis technique of ship infrared signature at sea. Journal of Ocean Engineering and Technology. 2009; 23(3): 53-58.

4. Ball RE, Calvano CN. Establishing the fundamentals of a sur- face ship survivability design discipline. Naval Engineers Jour- nal. 1994; 106(1): 71-74.

5. Driels M. Weaponeering: conventional weapon system effec- tiveness. 2nd ed. Virginia: AIAA Education Series; 2004. 1194 p.

6. Chung JH, Kwon JI. Survivability analysis of a vaval ship using the MOTISS Program(I): theoretical background. In: Proceed- ings of the Annual Autumn Meeting of Society of Naval Archi- tects of Korea; 2008 Nov 13-14; Chang-won, Korea, p. 330-339.

7. Kim KH, Kim JH, Choi TM, Kim YH, Cho DS. A study on a dynamic radar cross section analysis technique for a surface warship. Journal of Ocean Engineering and Technology. 2009; 23(6): 77-81.

8. Cho YJ. A study on measuring procedure and analysis techni- que of ship infrared signature at sea. Journal of Ocean Engineer- ing and Technology. 2009; 23(3): 53-58.

9. Nilubol O. Development of a combat aircraft operation and cost-effectiveness design methodology. PhD Thesis. Cranfield (UK): Cranfield College of Aeronautics, Cranfield University;

10. Michael OK. Radar systems, peak detection and tracking. 1st ed. Linacre(Oxford): Oxford Newnes; 2002. 363 p.

11. Kok SLW. Naval survivability and susceptibility reduction study surface ship. PhD Thesis. Naval Postgraduate School; 2012.

12. Fielding hp, Nilubol O. Integration of survivability assessment into combat aircraft design for operational effectiveness. In: Proceedings of the 24th International Congress of the Aeronau- tical Sciences; 2004 Aug 29-Sep 3; Yokohama, Japan; 1-10.

CHAPTER 2

A Mission-oriented Risk Assessment Methodology for Naval Vessel Fire Caused by Non-contact Explosions Using Bayesian Networks ☆

Jia Jia[a,b], Shouxiang Lu[a,*]

[a] State Key Laboratory of Fire Science, University of Science and Technology of China, Hefei 230026, China

[b] Bengbu Naval Petty Officer Academy, Bengbu 233012, China

ABSTRACT

In order to understand the fire damages on warship caused by non-contact explosion, a mission-oriented risk assessment methodology was developed in this paper. Firstly, an eight-step flow chart was proposed to analyze the fire effect on a warship, and the main factors, which determined the fire consequences, were summarized. Then based on the specialist marking methods and statistical data, the fire damage on warship could be calculated by using Bayesian Networks, which is a nice tool to assess the risk for complicated system. Finally, the methodology was applied a hypothetical ship with five compartments. The results show that fire intensity mainly depends on the fuel load and the fire damage depends on the function occupancy rate. Moreover, the result of sensitivity analysis indicates that fire damage could be effectively reduced by some ways, such as decrease leakage caused by shock, shorten the response time of fire detection and so on.

KEYWORDS

Mission-oriented; Fire risk assessment; Bayesian networks; Naval vessel; Non-contact explosions

1. INTRODUCTION

In wartime naval vessels would be hit by contact explosion and non-contact explosion. Although non-contact explosion does not directly hit naval vessel, it can cause fire by shock wave. As one of the most serious disasters, the ship fire attracts many attentions.

As to commercial ship, International Maritime Organization (IMO) approved a structured and systematic methodology for assessing risk which was named Formal Safety Assessment (FSA) [1]. This method was initially used in IMO rule- making process, and then for ship fire risk assessment [2]. FSA provides a relatively standard assessment process which focuses on fire frequency reduction, but it is not intended to cover risk control options for consequence reduction.

However, warship emphasizes govern severity of a fire rather than probability of accidental ignition [3] and FSA is not proper for warship. Table 1 shows some fire risk assessment methods focusing on fire severity reduction for naval vessels. A Ship Fire Safety Engineering Method (SFSEM) was developed from Engineering Method for Building Fire Safety by U.S. Coast Guard Research and Development Center. SFSEM is a probabilistic-based structured and comprehensive fire risk analysis methodology which treats all types of surface ships as a fire safety system [4]. All relevant aspects of fire safety are considered, including the growth and spread of fire, the effectiveness of passive and active fire protection measures [5]. Moreover, NKF Engineering Associates developed a Fire Loading and Self-inflicted Fire Risk Analysis for a U.S. Navy Destroyer design. This method aimed at achieving a balanced ship design with respect to fire hazards, compartment value and fire protection [6]. The method was then improved by Materials Research laboratory of Australia, and the assessment results were demonstrated in an easily understood set of colored charts [3].

Although two above methods focused on fire consequence (severity) rather than fire frequency (probability), the attentions were still paid on fires caused by self-inflicting in peacetime rather than by weapon attacks in wartime. In wartime, due to non-contact explosion attack, the fire ignition probability would greatly increase, the occurrences of fuel spills and hull crack would change combustion circumstance, and the damage of suppression system would reduce its efficiency. Compared to the condition in peace time, the wartime state would be more severe.

The purpose of this paper is to develop a Mission-oriented Risk Assessment methodology for naval vessel fire caused by Non-contact Explosions (MRANE). Firstly, an eight-step flow chart was given to introduce the evolution of fire and its effect on the ship. Secondly, the influence factors were figured out, and an index system was constructed in a hierarchy structure. Then based on Bayesian Networks, the fire damage on the warship could be calculated, and this method was used to study a case in the end.

Table 1. Application areas of different fire risk assessment methods

Fire cause	Methods focusing on fire frequency/probability reduction (merchant ship)	Methods focusing on fire consequence/severity reduction (naval vessel)
Internal cause (self-inflicting)	FSA	SFSEM/ Self-inflicted Fire Risk Analysis
External cause (weapon attacks)	—	MRANE (present work)

2. METHODOLOGY OF RISK ASSESSMENT

2.1 Assessment flow

For naval vessels, the fire risk caused by non-contact explosion can be analyzed as following the flow chart shown in Fig. 1.

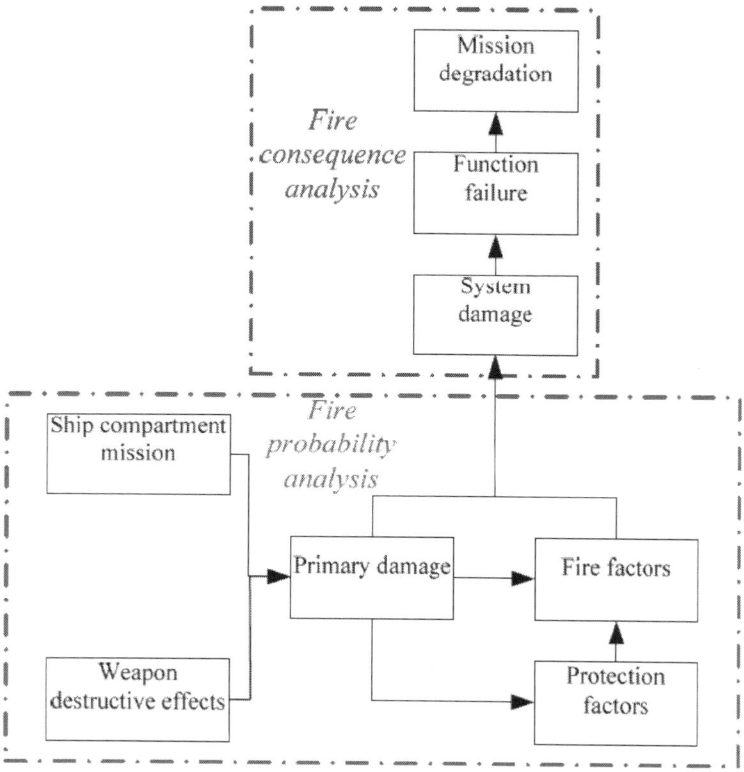

Figure 1. Flow chart of MRANE.

STEP 1: SHIP COMPARTMENT MISSION ANALYSIS

Assessment Goal:

The fire risk assessment goal is quite different for warship and merchant ships. For naval vessels, the goal of fire risk assessment is to evaluate the ship's residual ability to continue the assigned military mission after a fire caused by weapon attacks. So it can be called a mission-oriented assessment method. Among all of missions of naval vessels, the most important one is combat mission, which is concentrated in this paper.

Assessment scope:

Compartment is the basic unit of a ship. In this paper we tried to develop a method to assess the degradation of a compartment ability to perform combat mission after a fire caused by shock of non-contact explosions. Then the fire risk of each compartment in a ship could be evaluated, and the total fire risk of the ship could be obtained.

Assessment levels:

The function of compartment depends on the systems in the compartment. Thus, the assessment could be divided into three levels, *i.e.* mission level, function level and system level. As shown in Fig. 2, mission level is supported by function level which mainly includes fighting, moving and floating. The weight of each function varies with change of the mission. Function level is supported by system level which mainly includes fighting system, electric system, moving system and floating system.

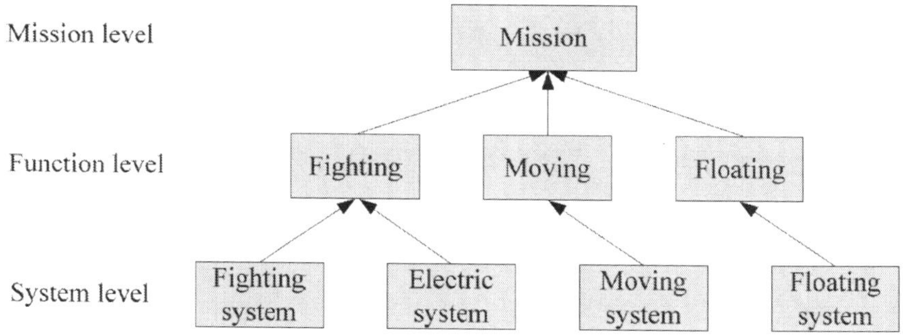

Figure 2. Assessment levels.

STEP 2: WEAPON DESTRUCTIVE EFFECTS ANALYSIS

In general, if the class and mission of a warship were known, the type of weapon which would attack the ship is determined. Then the hit probability and destructive can be obtained. This paper concentrates on non-contact explosions, thus primary damages are mainly caused by shock waves, and fire and flooding could induce secondary damages. In this paper, we just care about the fire risk.

STEP 3: PRIMARY DAMAGE ANALYSIS

In this paper the ship survivability is considered after a given shock loading and the shock damages on system functions: Hull cracks

System damages, such as failure of suppression systems and degradation of fighting systems Electric short circuit

Pipeline break

STEP 4: FIRE FACTORS ANALYSIS

Fire factors include ignition, fuel load and ventilation condition. The ignition may be caused by failure of system or weapon attacks, and the system failure could be ignored because of its low probability in wartime. So in this paper just the ignition by shock accident is considered. Fuel load includes the original fuel in the compartment and the fuel leakages from broken pipelines. Usually, it can be considered that the fire duration might to be proportional to the fuel load and fire intensity is proportional to the ventilation.

STEP 5: PROTECTION FACTORS ANALYSIS

As in FSA, active and passive fire protection measures are needed to control and extinguish fire. Active fire protection measures mainly include fire detection (auto and manual), fire suppression (auto and manual), firefighting training and management, shut down the ventilation and so on. Passive fire protection factors include the use of fire insulation on bulkheads and decks, fire tight cabling and piping bulkhead penetrations, and fire dampers in ventilation ducts. However, it should be noted that protection measures would fall failure partly or entirely under weapon shocks[7].

STEP 6: SYSTEM DAMAGE ANALYSIS

The main changes for the fire caused by the shock wave can be concluded as follows: The probability of ignition greatly increases

Fuel leaks from broken pipelines

Hull cracks change ventilation condition Efficiency of firefighting system decrease Step 7: Function failure analysis

Once the system damages have been evaluated, the function failure can be determined.

STEP 8: MISSION DEGRADATION ANALYSIS

Finally, if we take account of all these function failures, the mission degradation of the compartment can be achieved.

2.2 Construction of index system

Based on the above analysis, an index system of MRANE is constructed, as shown in Fig. 3. The index system is a set of related factors referred in the assessment flow, and is arranged in hierarchical structure. The top of this index system is ship mission, which depends on function factors, hazard factors and protection factors. Each of these three factors also includes some sub-factors. Finally, a factors tree is produced, and it can clearly illustrate the relationship among these factors.

2.3 Modeling by Bayesian network

BASIC THEORY OF BAYESIAN NETWORKS

Compared to the traditional risk assessment techniques, e.g. fault tree and event tree, Bayesian Networks (BN) can express complex net system and is a better analysis of a dependable system [8]. The use of BN modeling risk assessment had been suggested by Japan in 2006 [9], and was formally adapted in FSA by IMO in 2007 [10]. As a powerful modeling tool, BN has been successfully used to evaluate ship risk in many aspects, including grounding, collision and fire [9, 11-13].

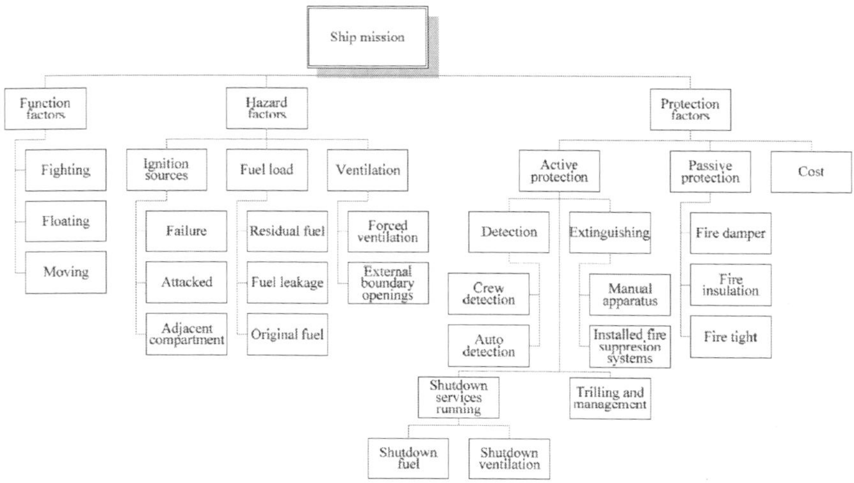

Figure 3. Index system of MRANE.

BN is a causal network that enables a graphical representation of causal relations between different parameters. It can be expressed as following:

$$BN = << V, E >, P > \quad (1)$$

V denotes a set of nodes, which can represent random variables. Each variable has a number of states.

E denotes a set of directed links, which denotes the causal relationship among the nodes. For example, a directed link $E = < V_i \quad V_j >$ denotes that node V_i has an effect on node V_j. So node V_i is called the parent of node V_j, and node V_j is called the child of V_i. The node without parents is called root node.

$< V, E >$ can identify a qualitative part of a risk model, while P can present its quantitative part. P describes how the states of a node depend on the parents of the node, *i.e.* a conditional probability table (abbreviated CPT).

FIRE RISK ASSESSMENT MODEL FOR COMPARTMENT

In order to build a fire risk assessment model for naval vessels shocked by non-contact explosions, some limitations and hypotheses are given:

- The ship is shocked by non-contact explosion when performing a combat mission. The ignition is caused by weapon attack.
- The functions supporting mission only include fighting and moving, while flooding is ignored. The effect of electric failure on fighting system and moving system is not considered.
- The activity of crew is not considered.
- The cost of fire protection is not considered.
- Compartments of the ship are considered independent, and no fire spread from compartment to compartment. So the passive protection system is assumed 100% effective.

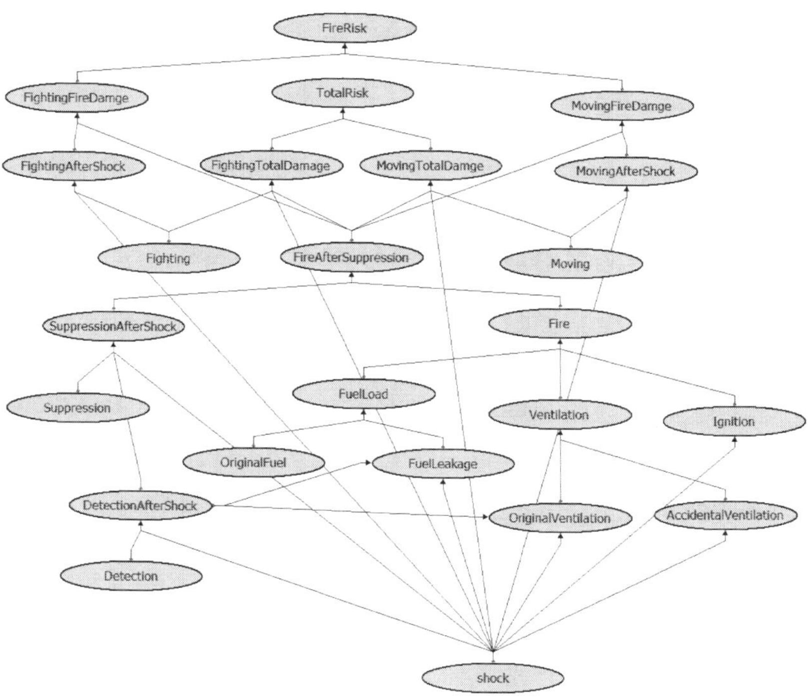

Figure 4. BN model of MRANE.

Then the BN model of MRANE can be constructed following the steps bellow.

Firstly, the nodes of the model should be determined. According to the eight-step analysis and the index system constructed above, a number of fire risk parameters have been achieved, as shown in Fig. 3. Based on the hypotheses above, some parameters can be ignored. Finally, 24 parameters are selected and each parameter is treated as a node (V_i) of BN model. All the nodes are listed in Table 2.

Secondly, the relationship among the nodes should be determined. Based on the eight-step analysis as well as the assumptions above, the relationship among the nodes are determined as Fig. 4.

Finally, the probability distribution of each node should be determined. Each node of the BN model is a binary-state random variable, *i.e.* YES or NO. The YES state affirms the occurrence of the event or operational status of the system represented by the node and NO indicates the opposite state. The definition of the YES state of each node can be seen in Table 2. The probability of a node in one state can be obtained in two ways, *i.e.* statistics and experts.

3. CASE STUDY AND ANALYSIS

3.1 Case description

Figure 5 shows a hypothetical frigate class ship with 5 compartments, which is represented by C1, C2, C3, C4 and C5.

The subdivision of fighting system, moving system, fuel loading and other main input data are given in Table 3.

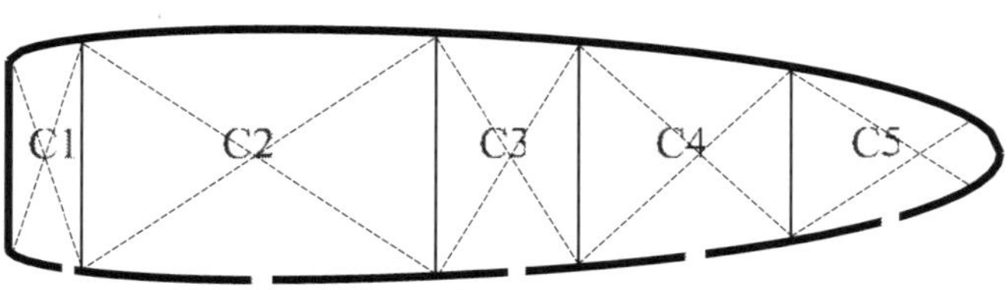

Figure 5. Compartment subdivision of a hypothetical frigate class ship.

Table 2. Nodes description of MRANE

No.	Name of node V_i	Definition of the YES state of V_i	Parent Node(s)
1	*Shock*	Ship is hit by non-contact explosion.	—
2	*Original Fuel*	The fuel load in a compartment before attack can sustain a fire which can destroy the whole compartment.	—
3	*Fighting*	The whole fighting system is in the compartment.	—
4	*Moving*	The whole moving system is in the compartment.	—
5	*Detection*	The fire can be detected as soon as it occurs.	—
6	*Suppression*	The suppression system can extinguish the fire in a compartment before shock.	—
7	*Detection After Shock*	After shock the detection system can detect fire as soon as fire occurs.	*Shock, Detection*
8	*Suppression After Shock*	The suppression system can extinguish the fire in a compartment after shock.	*Shock, Suppression, Detection After Shock*
9	*Fuel Leakage*	The fuel load of leakage can sustain a fire which can destroy the whole compartment.	*Shock, Detection After Shock*
10	*Fuel Load*	The sum of *Original Fuel* and *Fuel Leakage*.	*Original Fuel, Fuel Leakage*
11	*Original Ventilation*	The machinery ventilation can supply sufficient oxygen.	*Shock, Detection After Shock*
12	*Accidental Ventilation*	Ventilation caused by hull cracks can supply sufficient oxygen.	*Shock*
13	*Ventilation*	The sun of *Original Ventilation* and *Accidental Ventilation*.	*Original Ventilation, Accidental Ventilation*
14	*Ignition*	There is an ignition in the compartment.	*Shock*
15	*Fire*	There is a fire which can destroy the whole compartment.	*Fuel Load, Ventilation, Ignition*
16	*Fire After Suppression*	There is a fire which can destroy the whole compartment after suppression.	*Suppression After Shock, Fire*
17	*Fighting Total Damage*	The fighting function is lost caused by shock and fire.	*Fighting, Fire After Suppression, Shock*
18	*Moving Total Damage*	The moving function is lost caused by shock and fire.	*Moving, Fire After Suppression, Shock*
19	*Total Risk*	The sum of *Fighting Total Damage* and *Moving Total Damage*.	*Fighting Total Damage, Moving Total Damage*
20	*Fighting After Shock*	The fighting system still works after shock.	*Fighting, Shock*
21	*Moving After Shock*	The moving system still works after shock.	*Moving, Shock*
22	*Fighting Fire Damage*	The fighting function is lost caused by fire.	*Fighting After Shock, Fire After Suppression*
23	*Moving Fire Damage*	The moving function is lost caused by fire.	*Moving After Shock, Fire After Suppression*
24	*Fire Risk*	The sum of *Fighting Fire Damage* and *Moving Fire Damage*.	*Fighting Fire Damage, Moving Fire Damage*

4. RESULTS AND DISCUSSION

Applying the BN model built above to this case, the assessment results can be achieved in probabilistic forms. Fig. 6 (a) presents the fire probability of each compartment, which means how intense the fire will be. It can be seen that the fire probability can be ranked from high to low as C2, C5, C4, C3, C1, whether before or after suppression. It is mainly because of the quantity of fuel load in each compartment.

Table 3. Main input data of a hypothetical frigate class ship

Compartment No.	C1	C2	C3	C4	C5
P{*Fighting*=YES}	0	0.3	0.2	0.4	0.1
P{*Moving*=YES}	0	0.6	0	0.4	0
P{*Original Fuel*=YES}	0.1	0.6	0.2	0.3	0.4
P{*Shock*=YES}	0.6	0.6	0.6	0.6	0.6

Figure 6 (b) presents the fire risk (fire consequence) and total risk of each compartment. It ranks C2, C4, C3, C5, and C1 from high to low. This result disagrees with the fire intensity in Fig. 6 (a). This probably because that the compartment fire risk is not only impacted by fire intensity but also by the occupancy rate of fighting and moving in the compartment. For example, although its fire probability is 14.48, the fire risk of compartment C1 is zero, because there is no fighting or moving system in it. Moreover, from Fig. 6 (b), it can be concluded that total risk (the sum of shock damage and fire damage) is higher than fire risk (just fire damage) in a compartment, because total risk includes both fire and shock damage.

(a) 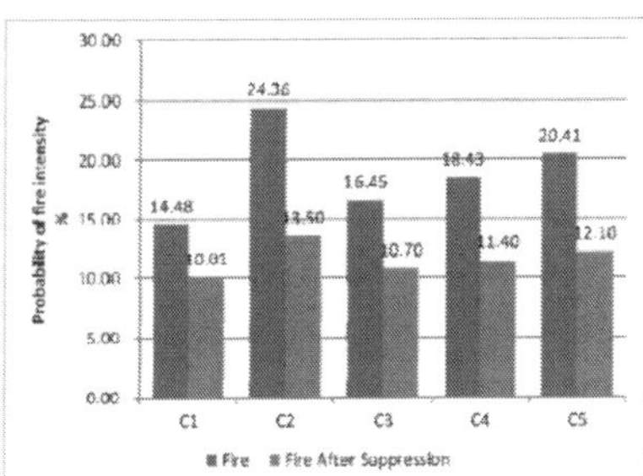

(b) 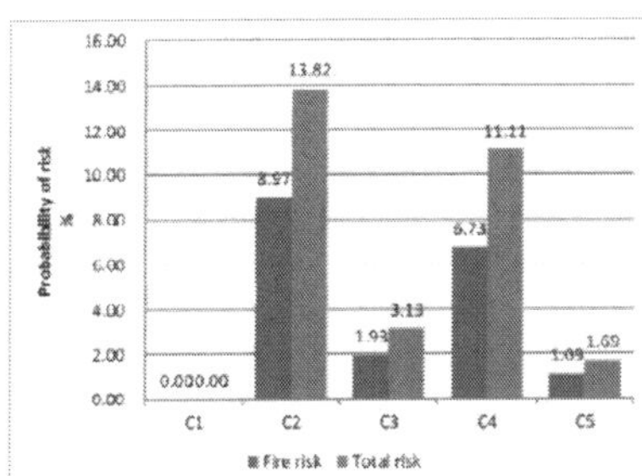

Figure 6. (a) The probability of nodes fire and fire after suppression. (b) the probability of nodes fire risk and total risk.

This paper focuses on how to reduce fire risk rather than shock damage. In order to reduce fire risk, it is necessary to analyze the sensitivity of fire risk. The sensitivity analysis equation is expressed as:

$$S_{FireRisk} = \left| \frac{P(FireRisk|V = \text{YES}) - P(FireRisk|V = \text{NO})}{P(V = \text{YES}) - P(V = \text{NO})} \right| \tag{2}$$

(a) 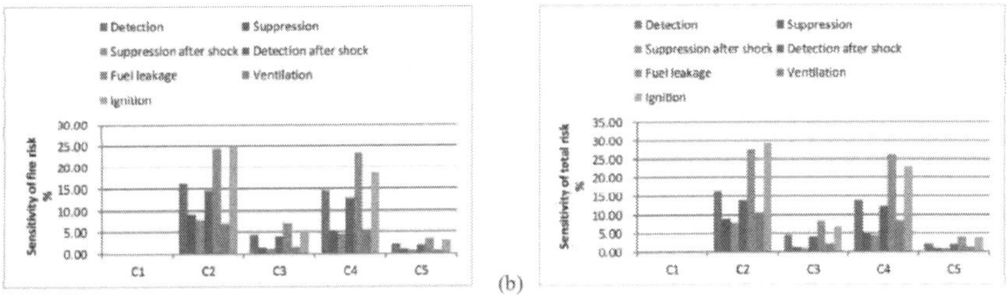 (b)

Figure 7. Sensitivity analysis of (a) fire risk. (b) total risk.

Figure 7 (a) presents the result of fire risk sensitivity analysis in each compartment. It is found that *Fire Risk* is highly influenced by *Fuel Leakage* and *Ignition*, moderately by *Detection* and *Detection After shock* and little by *Suppression*, *Suppression After Shock*, *Ventilation.* Furthermore, the result of total risk sensitivity analysis in each compartment basically corresponds with Fig. 7 (a).

According to Fig. 7 (a) and Fig. 7 (b), both fire risk and total risk can be reduced effectively:

- Decrease fuel leakages by enhancing shock resistance of pipelines.
- Decrease occurrence of electric arc, spark caused by shock.
- Shorten the detection response time and improve the survivability of detection system after shock.

5. CONCLUSIONS

When a naval vessel is attacked by non-contact explosion, the fire scenario would be great different from ship fire in peace time. Thus, a mission-oriented fire risk assessment methodology for naval vessel shocked by non-contact explosions (MRANE) was developed in this paper.

The key points of MRANE were three

- parts: An eight-step flow chart
- An index system in a hierarchy structure A BN model.

The flow chart is a mission-oriented assessment procedure, which makes it more standard and more systematic to evaluate fire risk of naval vessels. It can be applied to naval vessels not only under combat mission but also under other missions. The index system presents the effect factors of warship fire, which should be noticed in fire protection design. The BN model is a quantitative risk assessment model, which can simulate complex problems, such as ship fires, more precisely. The MRANE can be applied to naval vessel designs, and the assessment results can show clearly which compartment is overprotected and which is lack of protection. Thus, an optimized design can be achieved by balancing fire protection factors and other influence factors among compartments. Because MRANE is a mission-oriented method, finally it can promote the ship ability to continue combat mission in the circumstance of fire caused by non-contact explosions.

ACKNOWLEDGEMENTS

The present work was supported by National Natural Science Foundation of China with Project No. 50976109

REFERENCES

1. IMO, 2002. Guidelines for Formal Safety Assessment (FSA) for Use in the IMO Rule-Making Process. MSC/Circ. 1023.

2. Charchalis, A., Czy , S., Trial Application of the Formal Fire Risk Assessment to Fire Prevention in Ship Engine Rooms. Available from: http://www.polishcimac.pl/Papers2/2011/006. pdf.

3. Kennett, S. R., Lambrineas, P., Suendermann, B., Woodward, R. L., 1991. A Fire Risk Assessment Methodology for Naval Vessels. In: Conference, F.S. (Ed.).

4. Sprague, C., Richards, R., Blanchard, L., 1992. A Methodology for Evaluation of Ship Fire Safety, Naval Engineers Journal 104, p. 104.

5. Chester, M., Sprague, P. E., Brian, L. D., 1996. Theoretical Basis of the Ship Fire Safety Engineering Methodology, U.S. Coast Guard Research & Development Center.

6. Grzcszkiewicz, M. T., Pride, T. S., Davis, W., 1984. Fire Risk Assessment. Industry, Government, Education Design for the Future, p. 791. [7] IMO, 2009. Guidelines for Measures to Prevent Fires in Engine-rooms. MSC.1/Circ. 1321.

7. Bobbioa, A., Portinalea, L., Minichinob, M., Ciancamer lab, E., 2001. Improving the Analysis of Dependable Systems by Mapping Fault Trees into Bayesian Networks, Reliability Engineering & System Safety 71, p. 249.

8. IMO, 2006. Consideration on Utilization of Bayesian Network at Step 3 of FSA Evaluation of the Effect of ECDIS, ENC and Track control by using Bayesian network.MSC 81/18/1.

9. IMO, 2007. Consolidated Text of the Guidelines for Formal Safety Assessment (FSA) for Use in the IMO Rule-Making Process (MSC/Circ.1023 MEPC/Circ.392).MSC 83/INF. 2.

10. Passenger Hip Safety: Effective Voyage Planning for Passenger Ships, 2005. Formal Safety Assessment Large Passenger Ships Navigation. Sub Committee on Safety of Navigation, 50th Session, http://research.dnv.com/skj/FSALPS/ANNEXII.pdf

11. Friis-Hansen, A., 2000. Bayesian Networks as a Decision Support Tool in Marine Applications. Department of Naval Architecture and Offshore Engineering, Technical University of Denmark. PhD thesis.

12. Trucco, P., Cagno, E., Ruggeri, F., Grande, O., 2008. A Bayesian Belief Network Modelling of Organisational Factors in Risk Analysis: A Case Study in Maritime Transportation, Reliability Engineering & System Safety 93, p. 845.

CHAPTER 3

A Study on the Estimation Method of Risk Based Area for Jetty Safety Monitoring

Byeong-Wook Nam, Kyung-Ho Lee *, Jung-Min Lee, Dae-Seok Kim, Bon-Jae Ku, Runqi Li and Seong-Sang Yu

Department of Naval Architecture and Ocean Engineering, Inha University, 100 Inha-Ro, Nam-Gu, Incheon 22212, Korea

ABSTRACT

Recently, the importance of safety-monitoring systems was highlighted by the unprecedented collision between a ship and a jetty in Yeosu. Accordingly, in this study, we introduce the concept of risk based area and develop a methodology for a jetty safety-monitoring system. By calculating the risk based areas for a ship and a jetty, the risk of collision was evaluated. To calculate the risk based areas, we employed an automatic identification system for the ship, stopping-distance equations, and the regulation velocity near the jetty. In this paper, we suggest a risk calculation method for jetty safety monitoring that can determine the collision probability in real time and predict collisions using the amount of overlap between the two calculated risk based areas. A test was conducted at a jetty control center at GS Caltex, and the effectiveness of the proposed risk calculation method was verified. The method is currently applied to the jetty-monitoring system at GS Caltex in Yeosu for the prevention of collisions.

KEYWORDS

jetty; ship; collision prediction; risk calculation; risk based area; terminal management; safety monitoring

1. INTRODUCTION

Although technologies that support sailing ships, such as electronic chart display information systems (ECDISs) and automatic identification systems (AISs), have been developed, various maritime accidents—including collision sinking, running aground, and fires—continue to occur. These accidents have a large impact on the marine environment and incur economic losses. There have been continual efforts to avoid such accidents. In Korea, ~16,000 reported maritime accidents occurred from 1981 to 2009. Approximately 23% of these accidents were caused by collisions [1]. Additionally, ~20% occurred at the harbor or near coastlines [2]. In international cases, a report on merchant ship accidents in UK, protection and indemnity club (P&I) analyzed ~6000 cases that caused over 100,000 dollars in damages; ~62% of these accidents were caused by human error [3]. In January 2014, an unprecedented collision occurred between a WUYISAN ship and a jetty at GS Caltex in Yeosu. The state after this collision accident of a WUYISAN ship and a jetty is shown in Figure 1. Owing to this accident, ~1000 barrels of oil leaked, causing enormous economic losses and environmental pollution. The accident was due to human error and occurred during an attempt to berth at the jetty while exceeding the regulation speed. Consequently, the importance of safety-monitoring systems, which have been widely demanded, has emerged.

In this paper, we suggest a risk calculation method for pier safety monitoring and accident avoidance. The method identifies risk based areas for sailing ships and jetties, assesses the risks in these areas, and judges potential collisions. AIS data, principal particulars, Akira's equation, and other factors were used to compose a module for calculating the risk based areas. We developed a prototype and applied it to the pier safety-monitoring system. The effectiveness of the method was verified using the prototype.

Figure 1. Photographs taken after the collision between the ship WUYISAN and the oil pier of GS Caltex Co. (Yeosu, South Korea).

2. RELATED WORKS

Interest in safety monitoring has recently increased. Studies on maritime safety have been performed at pier control centers, as well as shipyards, laboratories, and universities. In particular, collision avoidance—the topic of this paper—has been addressed by many authors [4,5,6,7,8,9,10,11,12,13,14,15]. Researchers have studied systems to avoid collisions between ships by predicting collisions using fuzzy theories [4,5,6,7,8,9,10], introducing the concept of the ship domain [11,12,13], and improving existing systems [14,15].

Fuzzy inference, which was introduced by Zadiah, obtains a result corresponding to the input, according to the fuzzy rules [16]. This method has the advantage of describing expert knowledge intuitively. Factors related to the operation of the ship are employed in collision prediction using fuzzy inference. A collision-avoidance system using Mamdani-type fuzzy inference to predict ship collisions was introduced by Perera et al. [4]. Andrzej et al. suggested a collision-avoidance system that calculated the change in the prediction time and route of the ships [5]. Later, Yalei et al. developed a system for avoiding and preventing collisions between ships by using the time of the closest point of approach (TCPA), the distance of the closest point of approach (DCPA), and encounter angle [6]. Bukhari et al. devised a method for calculating the degree of collision risk using the DCPA, TCPA, and vessel traffic system (VTS) [7]. In addition to simple marine collisions, Simsir et al. proposed an artificial neural network using the Levenberg–Marquardt learning algorithm to prevent collisions when ships passed through a narrow strait [8]. A collision-avoidance system based on fuzzy rules and an ordered probit regression method was introduced by Chin et al. [9]. This system makes it easier to understand the collision risk by using a framework based on a probabilistic risk assessment model. The avoidance action can be defined at the appropriate level by using this system. Recently, a framework for risk-informed maritime collision avoidance system (CAS) was proposed by Goerlandt et al. [10]. This framework was developed to understand ship–ship encounter processes and their relation to collision risk.

The ship domain, which represents the virtual area of the ship, was introduced by Fujii [11]. It supports the avoidance of collisions by keeping the clean area from entering the other ships. The ship domain was later redefined by Goodwin [12]. Static objects, including ships, were added as detection targets in the domain. In addition, several researchers have proposed various methods employing the definition of the shape and size of the domain to increase the safety of the collision-avoidance support system. In particular, Mazaheri et al. defined a domain that is not limited to the collision and can be applied to the running-aground scenario [13]. The domain was expanded downdraft, and a system was developed to avoid collisions as well as running aground.

A system supporting collision avoidance was developed by improving existing systems, including navigation equipment such as electronic chart display & information system (ECDIS) [14,15]. A collision-avoidance system developed by Kim et al. based on ECDIS was linked with various navigation communication equipment and calculated the collision risk [14]. The system creates a collision avoidance route for when the risk is greater than the reference value. Kim et al. changed the existing system into an integrated linkage-monitoring system that recognizes and provides warnings of dangerous situations and indicates a method for reaching a safe berth [15]. They tried to enhance the safety monitoring by improving the existing system.

The introduced systems detect collisions between ships as well as instances of running aground, and they support avoidance. However, they do not prevent collisions between a ship and a jetty, such as the WUYISAN accident. Therefore, we introduce the concept of the risk based area, which employs the concept of the marine domain. We suggest a method that predicts and avoids collisions between a ship and a jetty by assessing the collision risk using risk based areas.

3. METHOD FOR THE CALCULATION OF THE RISK BASED AREA

The risk based area is divided into those of the ship and jetty. Collisions between ships or between a ship and a jetty are predicted using these two areas. The calculation method for each risk based area is as follows.

3.1. Calculation of the Risk Based Area of a Ship

Researchers have proposed various methods employing the definition of the shape and size of the domain. Collision prediction by using ship-maneuvering data was first proposed by Montewka et al. [17]. They predicted a collision between two ships by calculating the shortest distance between them. Later, this concept was further extended. Göhler defined the expectation area according to the route of the ship [18]. In particular, Baldauf et al. defined an umbrella-shaped ship domain by using the turning and stopping performances [19]. In this paper, the risk based area of a ship is calculated by using the shortest stopping distance and the turning stopping distance of the ship. The risk based area of the ship is pentagonal, with five vertices that are determined according to the shortest stopped location, turning stopped location, and stern.

3.1.1. Shortest Stopping Distance of a Ship

Because a ship is a dynamic object, the velocity of the ship, the shortest stopping distance, and the turning stopping distance are considered in calculating the risk based area. The shortest stopping distance is the distance between the current ship location and the ship location when the hull is completely stopped, after starting full astern. The factors affecting the shortest stopping distance are as follows [20].

- Astern engine power: The stopping distance is inversely proportional to the magnitude of the astern power.
- Propeller type: The time required to go astern is shorter for ships with controllable pitch propellers than for those with fixed pitch propellers.
- Displacement: The time required to go astern is proportional to the displacement.
- Velocity: The stopping distance is proportional to the velocity before the engines go astern.
- Wind, hull form, ocean and tidal currents, etc.

There are two representative equations for calculating the shortest stopping distance: Knight's equation and Akira's equation [20]. Knight's equation represents the relationship between the astern stopping distance and time required and was obtained by the experience of Knight. Akira's equation represents the relationship between the full astern state and the velocity of the ship [20]. The two equations can be expressed as follows:

$$S = \frac{1}{5}vt \tag{1}$$

$$S = 0.0135\frac{v^2W}{Ta} \tag{2}$$

$$Ta = 0.095\frac{BHP}{v} \tag{3}$$

where S is the stopping distance (m), v is the velocity before the engines go astern (knot), t is the braking time (s), W is the displacement (ton), Ta is the astern power, and BHP is the brake horsepower.

Because Knight's equation calculates the distance through a test of the shortest stopping distance of the ship, Akira's equation—which can calculate the shortest stopping distance using the current velocity—was used in this study. Table 1 shows an example of the shortest stopping distance when a fully loaded very large crude carrier (VLCC) went full astern during navigation at a normal harbor speed of 12 knots.

Table 1. Relationship between the length of a ship and the shortest stopping distance for a fully loaded VLCC under sail at 12 knots.

Ship length (m)	Shortest Stopping Distance of Ship
210	10 times the length
250	13 times the length
300	15 times the length

The shortest stopping distance is calculated using Akira's equation. If the velocity is as low as 2.5 knots, the stopping distance is very small compared with the ship's length. Therefore, the shortest stopping distance is not calculated when the velocity is as low as 2.5 knots.

3.1.2. Turning Stopping Distance of a Ship

The turning stopping distance of a ship is the distance between the location where the braking started and the location where the ship stopped after turning into port or starboard, as illustrated in Figure 2. The turning stopping distance is expressed for two directions, in contrast to the shortest stopping distance. These two directions are based on the direction of the bow immediately before starting the turn and the turning direction. The stopping distance is calculated for each direction. The equations for calculating the port or starboard turning stopping distance are as follows:

$$\mathrm{F.S} = \frac{3}{4}\mathrm{S} \tag{4}$$

$$\mathrm{P.S} = \mathrm{S.S} = \frac{1}{2}\mathrm{S} \tag{5}$$

where F.S is the forward component of the turning stopping distance (m), S is the shortest stopping distance as calculated in the previous section (m), P.S is the lateral component of the turning stopping distance in the port direction (m), and S.S is the lateral component of the turning stopping distance in the starboard direction (m).

The forward component of the turning stopping distance is 0.75 times the shortest stopping distance, and the lateral component of the stopping distance is 0.5 times the shortest stopping distance [21]. For the uncertainty, the variable α was used. The stopping distance obtained by adding the calculated distance to α was used. The risk based area was drawn using the locations corresponding to the shortest stopping distance, the port or starboard turning stopping distances, and the end of the stern. The risk based area of the ship was pentagonal in shape, as shown in the shaded region of Figure 2.

3.2. Calculation of the Risk Based Area of a Jetty

In contrast to the risk based area of a ship, the risk based area of a jetty is set by consultation between the managers, including system operators, managers responsible for safety, and officials of relevant ministry to enact rules and laws. This area is determined in consideration of such factors as the size of the ship, the size of the jetty, the structure of the jetty, and the item of acquisition. Because it is impossible to directly avoid the jetty when the ship approaches, the area was divided into two—the dangerous area and the warning area—which allowed for flexible coping according to the situation. The International Maritime Organization (IMO) proposed a regulation velocity according to the distance from the jetty [22,23]. The regulation ship velocity and distance specified by the IMO Standards of Training, Certification, and Watchkeeping for Seafarers (STCW) rules are shown in Table 2 [22,23]. After the accident involving the Torry Canyon ship near the coastlines of France in March 1967, the STCW was formed by the IMO as an international convention for providing international standards regarding crew certification. This convention sets goals regarding the safety of ships, the prevention of marine pollution, and the improvement of crew qualifications, along with suggesting standards for training, certification, and watchkeeping [24].

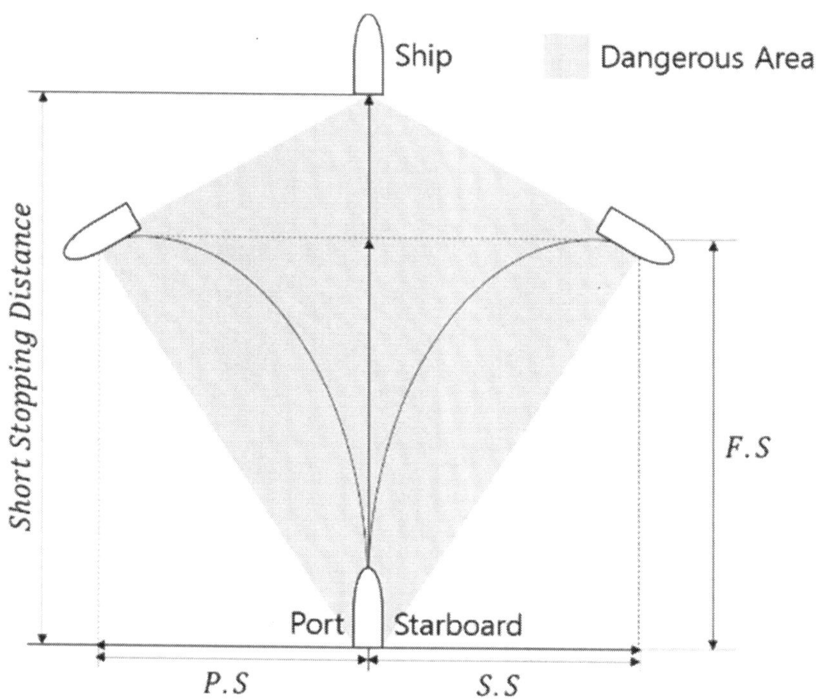

Figure 2. Set of risk based area and turning stopping distance of a ship.

Table 2. Regulations for ship velocity and distance according to the IMO STCW rules.

Distance	Deadweight Tonnage (DWT)		
	Below 10,000 DWT	Panamax, Cape	VLOC, ULOC
Within 1–2 L	2–3 knots	2–3 knots	2–3 knots
1 mile	7–8 knots	5–6 knots	4–5 knots
2 miles	13 knots	9 knots	7 knots
3 miles	Engine S/B	12 knots	9 knots
4 miles	Usual service condition	Engine S/B	11 knots
5 miles	-	Usual service condition	Engine S/B
notes	Approaching velocity 2–3 knots + 4 knots/mile	Approaching velocity 2–3 knots + 3 knots/mile	Approaching velocity 2–3 knots + 2 knots/mile

Engine S/B (standby): Before arriving in a port or departure from a port, we should standby the engine to meet the requirements of maneuvering operations in the port.

The risk based areas of the jetty, which are set according to various factors such as the weather and sea state, are shown in Figure 3. The risk based areas are rectangular or jetty-shaped. However, for calculating the turning route of a ship using the risk based area of the ship, a rectangular risk based area may be safer than a jetty-shaped one. These contents are covered concretely in the following risk calculation process.

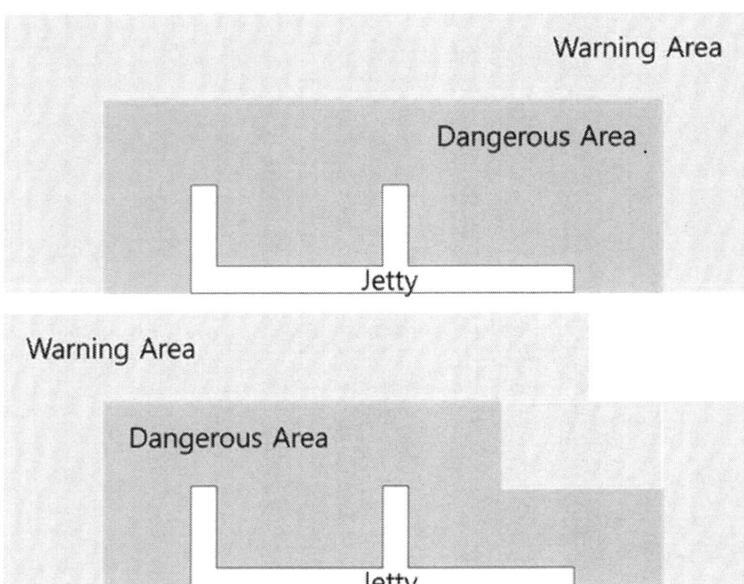

Figure 3. Set of risk based areas of a jetty (top and bottom figures show rectangular and jetty-shaped risk based areas, respectively).

The risk based area can also be expressed by applying the rules of the marine pilots association or IMO STCW according to the regulation speed. Risk based areas obtained using the rules of the Yeosu Marine Pilots Association, which are based on the IMO STCW rules, are shown in Figure 4. These are the risk based areas of the GS Caltex jetty in Yeosu.

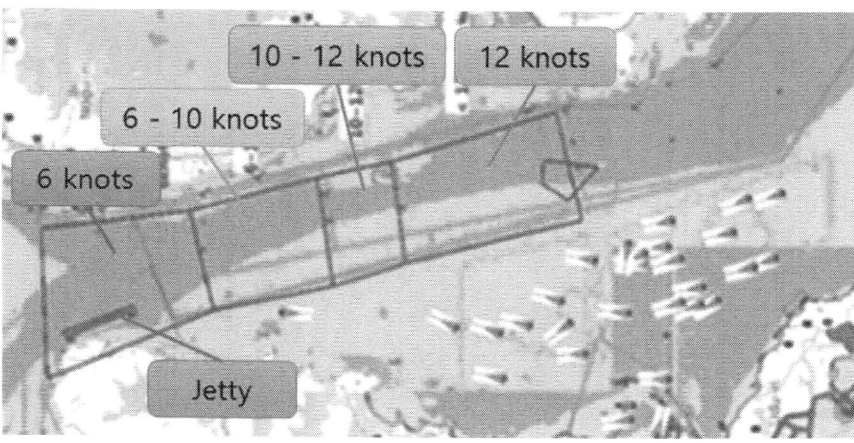

Figure 4. Set of risk based areas for a jetty according to the rules of the Yeosu Marine Pilots Association.

4. RISK CALCULATION PROCESS

Risk assessment is performed according to the process shown in Figure 5. First, the shortest stopping distance and turning stopping distance are calculated by using AIS data for ships obtained in real time. The risk based area for a ship is then calculated using the two stopping distances, and that for a jetty is calculated considering the relevant conditions. Second, the overlap between the risk based areas of the ship and jetty is calculated. Finally, collisions between the ship and jetty are predicted by assessing the size of the overlapping risk based area, and the results are displayed.

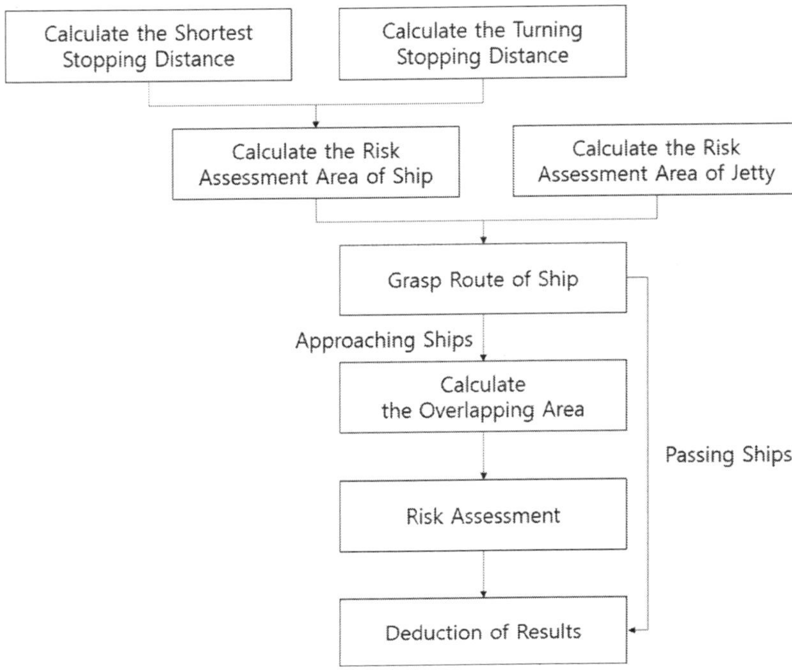

Figure 5. Risk calculation process.

Figure 6 shows a collision state according to the overlap between the risk based areas for the ship and jetty. The amount of overlap varied according to the shape of the risk based areas. The risk based area for the jetty was set as rectangular and the shape of a pier. However, if the risk based area for the jetty was set as jetty-shaped, as shown in Figure 6A, a situation in which the miss occurred in the collision prediction same as (c') in Figure 6A. In this case, because the collision warning shown in Figure 6A occurs later than that shown in Figure 6B, the probability of an accident is increased. Additionally, the response for the accident can occur a considerable amount of time after the accident, as in (c) in Figure 6B. Therefore, we set the risk based areas for the jetty as rectangular, as shown in Figure 6B.

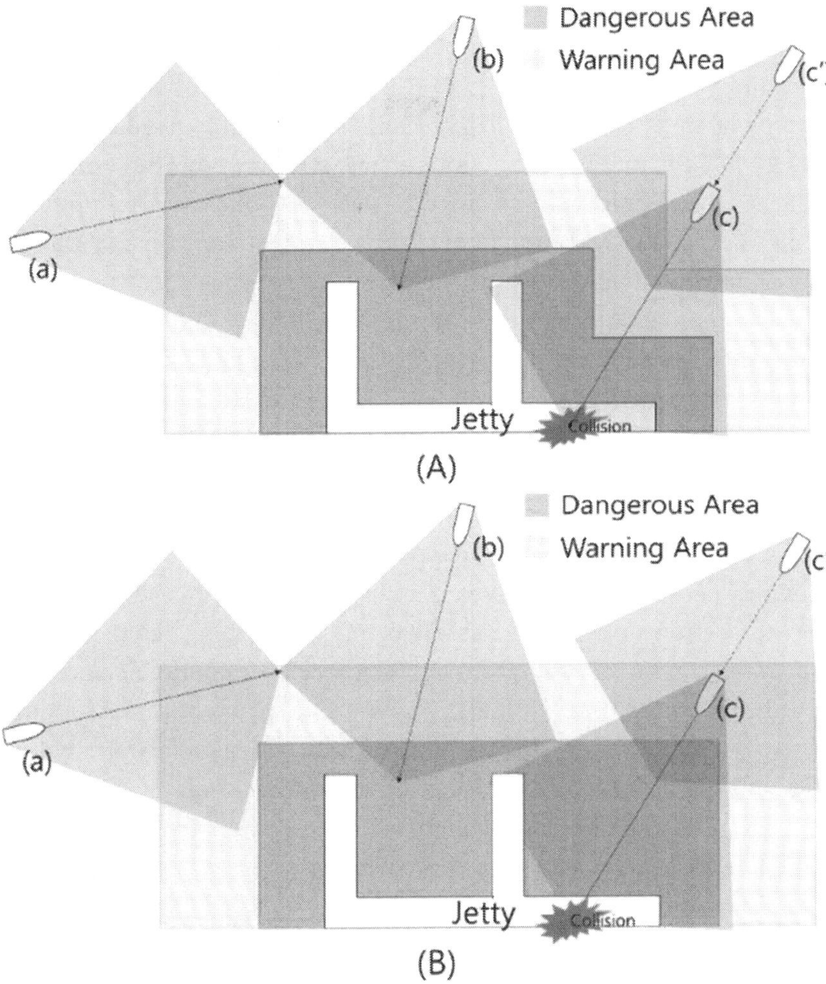

Figure 6. Risk assessment for the collision according to the route and location of a ship, jetty-shaped risk based areas (**A**); and rectangular risk based areas (**B**).

The risk is assessed by calculating area of overlap between the risk based areas of a ship and a jetty. Ships have a variety of purposes, including sailing, passing near a jetty, and approaching a jetty. Accordingly, if a system monitors dozens to hundreds of ships operating near a jetty and the collision warning alerts are abused, the system administrator may be confused. Therefore, we classified the ships by their direction of movement to avoid collision warning alert abuse. The ships were classified into two: those passing near the jetty instead of heading for the jetty, as indicated by (a) of Figure 6B; and those approaching the jetty, as indicated by (b) and (c') of Figure 6B. Risk assessment was performed for ships heading toward the jetty, excluding the ships passing near the jetty. If a collision warning alert about ships approaching the jetty occurs, the collision warning alert is similarly abused. Therefore, the velocity of the ship was used to avoid the abuse of the collision warning alert. Ships are subject to the regulation speed when operating near the jetty. The regulation speed follows the rules of the IMO STCW or the marine pilots association. The warning alert is for when ships exceeding the regulation speed approach the jetty. After the ships to be monitored were classified, the risk assessment was performed. The risk was assessed by judging the area of overlap between the risk based areas of the ship and the jetty. Because the overlap area depended on the risk based area of the jetty, the boundary of the risk based area of the ship—including the stopping point—was used to calculate the overlap area. Prior to performing a risk assessment, the turning boundary line connecting points of the port and starboard stopping distance was defined. Figure 7 presents a holistic view of the overlap-area assessment for clarity. The warning alert was set when the boundary of the risk based area for the ship touched that of the warning area of the jetty, as shown in (a') of Figure 7, or immediately before the boundary of the risk based area for the ship touched the boundary of the dangerous area for the jetty, as shown in (a) of Figure 7. The warning alert represents a low collision probability, but caution is required. The danger alert was set when the boundary of the risk based area for the ship touched the boundary of dangerous area for the jetty, as shown in (b') of Figure 7, or when the ship could turn in one direction (port or starboard) or avoid a collision by going full astern. The danger alert represents a high collision probability, but it is possible to prevent a collision through a quick response. Finally, when the boundaries of the risk based areas of a ship and a jetty intersect, the state is set to collision, as shown in (c) of Figure 7. That is, when the boundary of a jetty and a turning boundary line intersect and the boundary of a jetty and the shortest stopping distance line also intersect, a collision between the ship and the jetty occurs.

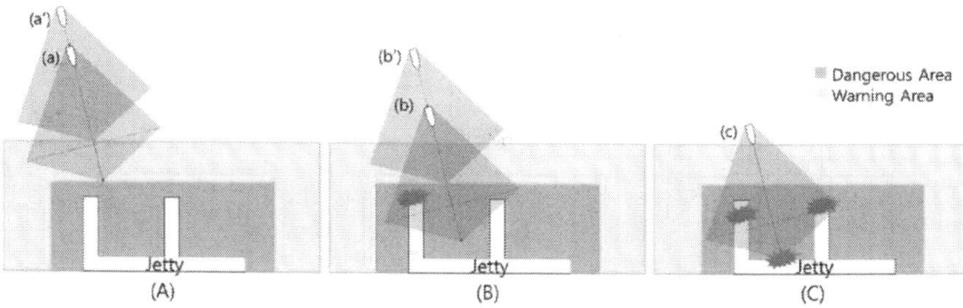

Figure 7. Risk assessment for a collision according to boundaries of risk based areas of a ship and a jetty; **(A)** represents the warning alert state, **(B)** represents the danger alert state, and **(C)** represents the collision state.

5. CONFIGURATION OF THE SAFETY-MONITORING SYSTEM FOR A JETTY

This paper suggests a jetty safety-monitoring system that consists of AIS for obtaining navigation information, a module for risk assessment, and a display for presenting the results. Figure 8 shows a jetty safety-monitoring system configuration applying the risk calculation method.

AISs are capable of two-way data communication between ships or between a ship and land. They automatically transmit and receive information related to the cargo, ship specification, and navigation information. One such system was adopted in December 2000 by the IMO. Through the maritime mobile service identity of ships and the control method using the navigation information, AISs have been used as a precautionary measure for preventing ship collisions by identifying the ships [25]. As shown in Table 3, the information provided by the AIS consists of static information, dynamic information, navigation information, and text communications [26]. In this study, data comprising the ship specification, location, and velocity was obtained from the AIS equipment. The risk based area of the ship and jetty were calculated according to the risk based area methodology described in Section 3 and Section 4 using the acquired data. Whether a collision could occur was then determined by evaluating the degree of overlap of each area through a risk calculation process. Ultimately, the result derived from the risk calculation method was applied to the pier safety-monitoring system and expressed on a jetty control monitor.

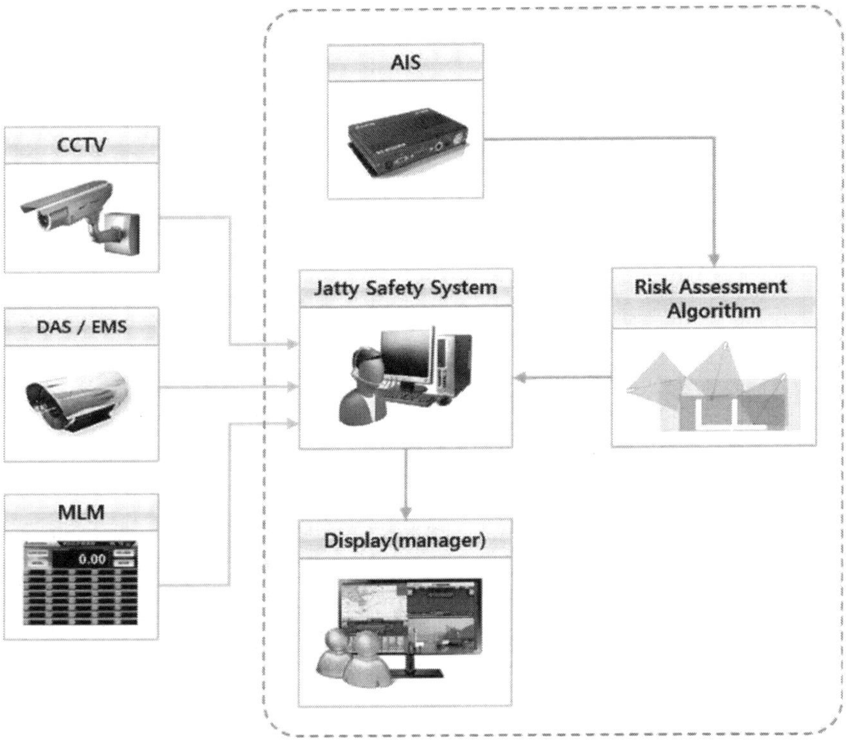

Figure 8. Configuration of a jetty safety-monitoring system using the risk calculation method.

Table 3. Information provided by AIS.

Division	Contents of Information
Static information	Name of ship, IMO registration number, call sign, kind of ship, ship specification (length, depth, width, *etc.*), location of antenna
Dynamic information	Ship location, course over the ground (COG), velocity, heading of ship, navigation state, rate of turn, angle of inclination
Navigation information	Ship draft, hazardous cargo, destination. estimated time of arrival, route planning
Notification letters	Important sailing or weather alerts

6. REALIZATION OF THE SAFETY-MONITORING SYSTEM FOR A JETTY

A prototype of the jetty safety-monitoring system applied to the risk calculation method is shown in Figure 9. The prototype was first applied to the second oil pier of GS Caltex; a map of the system was also implemented for the second oil pier. Pier traffic regulations and the jetty were considered in order to set up the risk based area for the jetty. As shown in Figure 10, the risk based area for the second oil pier was divided into four areas: 6 knots, 4–10 knots, 10–12 knots, and greater than 12 knots. The shortest stopping distance and turning stopping distance were calculated using the velocity and the direction of movement of the ship at the position acquired from the AIS. The risk based area calculated using the two stopping distances was overlaid on the map.

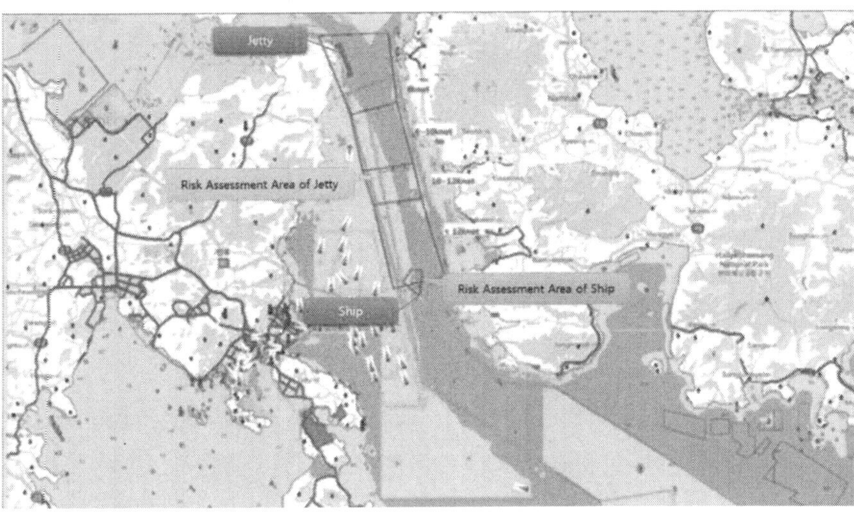

Figure 9. Map screen of the realized risk based area of the ship and jetty.

7. TEST AND VERIFICATION

The risk calculation method developed in this study was applied to the pier safety-monitoring system of the second GS Caltex oil pier. The test was conducted by targeting ships passing near the pier, approaching, and berthing. While testing the data for sailing ships, dozens to hundreds of ships passed near the pier. Accordingly, to determine the accuracy of the method, the test was conducted by intensively monitoring a ship whose purpose was to approach and berth at the pier. Figure 10 shows a screen that displays information about the ship as the object and monitors the movement of the object.

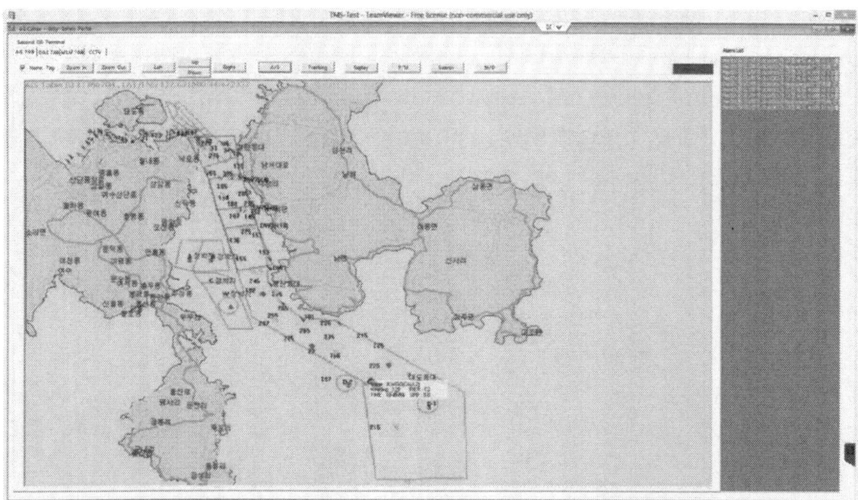

Figure 10. Map screen of the realized risk based area of the ship and jetty.

As the ship approached the pier, if it was located in the risk based area of the jetty set in accordance with the regulation velocity, or if a collision between the ship and the jetty was to be expected, a warning message was output. When a collision between the ship and the jetty was expected, the warning message shown in Figure 11 was displayed. This test was duplicated for many regions within the risk based area. The warnings are sent by a user. With a minimum of preventive measures to handle any warning generated by a system error, after the approval of the user at the jetty monitoring center, the warning system sends an alert.

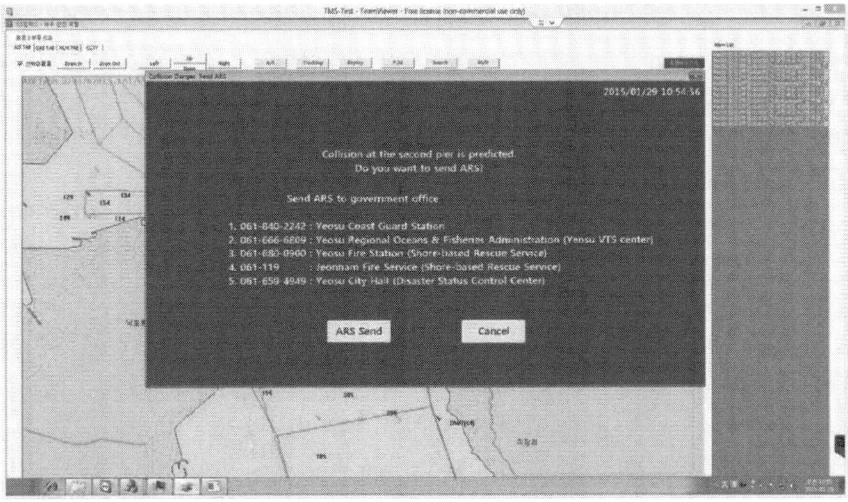

Figure 11. Collision warning message for the realized risk based area.

A prototype was developed using a risk calculation method and applied to the monitoring system of the second oil pier at GS Caltex. A test demonstrated the effectiveness of the risk calculation method for the pier safety-monitoring system. Compared with the existing system, the collision-prediction system introduced in this study improved the safety-monitoring method.

8. CONCLUSIONS

In this study, we introduced the concept of the risk based area, which is distinct from the methodology used by previous studies, and suggested a risk calculation method using this concept. The risk based area of ships and jetties was set using the shortest stopping distance, turning stopping distance, and AIS data of ships approaching the pier or passing near the pier. The risk calculation method calculates the overlapping area of risk based area through a risk calculation process. Finally, this method is used to determine whether or not a collision will occur between a ship and jetty. We developed a prototype using this method and applied the prototype to the second oil pier at GS Caltex. The effectiveness of the system was demonstrated by testing when ships are approaching and passing the pier. We confirmed an improvement in the collision-prediction capability of the jetty safety-monitoring system compared with existing systems. Because the method proposed in this paper was designed to be adaptable to the individual characteristics of a pier, it is expected to be easy to implement. Ultimately, applying this method to each jetty is expected to improve jetty safety monitoring.

ACKNOWLEDGMENTS

This work was supported by the Industrial Strategic Technology Development Program (10041790) funded by the Ministry of Knowledge Economy (MKE, Korea) and INHA research grant.

AUTHOR CONTRIBUTIONS

Byeong-Wook Nam, Jung-Min Lee, Dae-Soek Kim, Bon-Jae Ku, Runqi Li, and Seong-Sang Yu conducted the research and wrote the paper. Kyung-Ho Lee was the thesis supervisor. All authors provided guidance, revision, and feedback.

CONFLICTS OF INTEREST

The authors declare no conflicts of interest.

REFERENCES

1. Korea Maritime Institute Editorial department. Shipping Statistics Handbook 2010; Korea Maritime Institute: Busan, Korea, 2010; pp. 357–367.

2. Cho, D.O. A study on port safety management between Korea and Japan. Ocean Policy Res. **2003**, 18, 97–122.

3. Anderson, D.E.; Oberman, F.R.; Malone, T.B.; Baker, C.C. Influence of Human Engineering on Manning Levels and Human Performance. Navig. Eng. J. **1997**, 109, 67–76.

4. Perera, L.P.; Carvalho, J.P.; Soares, C.G. Autonomous guidance and navigation based on the COLREGs rules and regulations of collision avoidance. In Proceedings of the International Workshop on Advanced Ship Design for Pollution Prevention, Split, Croatia, 23–24 November 2009.

5. Lenart, A.S. Manoeuvring to required approach parameters—CPA distance and time. Annu. Navig. **1999**, 1, 99–108.

6. Ren, Y.; Mou, J.; Yan, Q.; Zhang, F. Study on Assessing Dynamic Risk of Ship Collision. In Proceedings of the Study on Assessing Dynamic Risk of Ship Collision (ICTIS 2011), Wuhan, China, 30 June–2 July 2011.

7. Bukhari, A.C.; Tusseyeva, I.; Lee, B.G.; Kim, Y.G. An intelligent real-time multi-vessel collision risk assessment system from VTS view point based on fuzzy inference system. Expert Syst. Appl. **2013**, 40, 1220–1230.

8. Simsir, U.; Amasyali, M.F.; Bal, M.; Celebi, U.B.; Ertugrul, S. Decision support system for collision avoidance of vessels. Appl. Soft Comput. **2014**, 25, 369–378.

9. Chin, H.C.; Debnath, A.K. Modeling perceived collision risk in port water navigation. Saf. Sci. **2009**, 47, 1410–1416.

10. Goerlandt, F.; Montewka, J.; Kuzmin, V.; Kujala, P. A risk-informed ship collision alert system: Framework and application. Saf. Sci. **2015**, 77, 182–204.

11. Fujii, Y.; Tanaka, K. Traffic Capacity. J. Navig. **1971**, 4, 543–552.

12. Goodwin, E.M. A statistical study of ship domains. J. Navig. **1975**, 28, 328–344.

13. Mazaheri, A.; Goelandt, F.; Kujala, P.; Montewka, J. A Decision Support Tool for VTS Centers to Detect Grounding Candidates. Int. J. Mar. Navig. Saf. Sea Transp. **2012**, 6, 337–343.

14. Kim, D.H.; Park, J.Y.; Lee, Y.S. A Study on the Development of Integrated Information System of Safety Management for Oil Tank Terminal; Korean Society of Marine Environment & Safety: Busan, Korea, 2014.

15. Kim, D.J.; Ahn, K.S.; Lee, T.I.; Kim, Y.W. Development of Collision Avoidance Supporting System Based on ECDIS; Korean Institute of Navigation and Port Research: Busan, Korea, 2013.

16. Lee, J.E. A Study on the Road Tunnel Lighting System Appied Fuzzy Reasoning and Dimming Control. Ph.D. Thesis, Hongik University, Seoul, Korea, 2015.

17. Baldauf, M.; Mehdi, R.; Deeb, H.; Shröder-Hinrichs, J.U.; Benedict, K.; Krüger, C.; Fischer, S.; Gluch, M. Manoeuvring areas to adapt ACAS for the maritime domain. Sci. J. Marit. Univ. Szczec. **2015**, 43, 39–47.

18. Montewka, J.; Goerlandt, F.; Kujala, P. Determination of collision criteria and causation factors appropriate to a model for estimating the probability of maritime accidents. Ocean Eng. **2012**, 40, 50–61.

19. Göhler, U.D. Estimation of Expectation Areas of Ships considering resistance changes, due to yaw angle and according to Model experiments. Schiffbauforschung **1983**, 4, 235–246.

20. Vessel Traffic Service Center. Available online: http://www.vtskorea.info/Board.do?flag=det&id =yeonjae_lecture&cPage=2&idx=51&limit=10&sc=SUBJECT&sv= &cg= (accessed on 26 July 2015).

21. Wang, H.; Wei, Z.; Wang, S.S.; Ow, C.S.; Ho, K.T.; Feng, B.J.M. A Vision-based Obstacle Detection System for Unmanned Surface Vehicle. In Proceedings of the 2011 IEEE Conference on Robotics, Automation and Mechatronics (RAM), Qingdao, China, 17–19 September 2011.

22. Incheon Port Marine Pilot's Association. Available online: http://www.incheonpilot.com/main/main.asp (accessed on 26 July 2015).

23. Pyeongtaek-Dangjin Port Marine Pilot's Association. Available online: http://www.ptpilot.co.kr/pilotage/pilotage_08_06_01.aspx (accessed on 26 July 2015).

24. Standards of Training, Certification & Watchkeeping. Available online: http://www.stcw.org/ (accessed on 25 August 2015).

25. Kim, C.J.; Song, J.U.; Kim, C.H. A Study on the Development of the Module for Remote-Controlling Aids to Navigation. J. Korean Navig. Port Res. **2002**, 26, 269–273.

26. Seo, M.H. Service Implementation for the AIS Information and Maritime Safety Information in the Internet. Master's Thesis, Mokpo National Maritime University, Mokpo, Korea, 2012.

CHAPTER 4

An Energy Management System of a Fuel Cell/Battery Hybrid Boat

Jingang Han [1,2,*], Jean-Frederic Charpentier [2] and Tianhao Tang [1]

[1] Department of Electrical Engineering, Shanghai Maritime University,1550 Haigang Ave, 201306 Shanghai, China

[2] French Naval Academy Research Institute, French Naval Academy, 29240 Brest Cedex 9, France

ABSTRACT

All-electric ships are now a standard offering for energy/propulsion systems in boats. In this context, integrating fuel cells (FCs) as power sources in hybrid energy systems can be an interesting solution because of their high efficiency and low emission. The energy management strategy for different power sources has a great influence on the fuel consumption, dynamic performance and service life of these power sources. This paper presents a hybrid FC/battery power system for a low power boat. The hybrid system consists of the association of a proton exchange membrane fuel cell (PEMFC) and battery bank. The mathematical models for the components of the hybrid system are presented. These models are implemented in Matlab/Simulink environment. Simulations allow analyzing the dynamic performance and power allocation according to a typical driving cycle. In this system, an efficient energy management system (EMS) based on operation states is proposed. This EMS strategy determines the operating point of each component of the system in order to maximize the system efficiency. Simulation results validate the adequacy of the hybrid power system and the proposed EMS for real ship driving cycles.

KEYWORDS

fuel cell (FC); hydrogen; battery; energy management system (EMS); hybrid power boat; driving cycle

1. INTRODUCTION

Emissions related to ship propulsion (CO_2, SO_x and NO_x) have a significant responsibility for environmental problems such as the greenhouse effect, acid rain and air pollution. It is reported that the CO_2 emissions from ships accounts for about 3%–5% of total CO_2 emissions in the world and corresponded to about 1 billion tons in 2010. It has been estimated by international maritime organization that, without any countermeasures, the carbon dioxide emission from ships will increase to 2.5 billion tons in 2050. Moreover, SO_2 emissions of ship are also a problem. It is estimated that it will increase by 10%–20% in 2012, which corresponds to 5.2% of the total sulphate burden in the world [1].

Although shipping emissions are not covered by the Kyoto Agreement, stricter regulations have been developed in different countries. As an example, European Union (EU) has implemented stricter regulation to control the SO_2 emissions in particularly sensitive sea areas, including the Western European Waters and the Baltic Sea areas. In those areas, sulphur content in the fuel used in ship should not exceed 0.1%. However, this content is more than 1% in most marine fuels [1,2]. This is why the development of alternative green energy sources has become a critical issue for ship propulsion. However, the high cost, the relatively low energy and power level and the technical barriers of alternative power systems such as fuel cell (FC) or other renewable energy systems limit their applications in ships.

One of the most promising technologies which can be used in a ship is FC system (FCS). There are several types of FCs [3]: the proton exchange membrane fuel cells (PEMFC), alkaline fuel cells (AFC), phosphoric acid fuel cells (PAFC), molten carbonate fuel cells (MCFC), solid oxide fuel cells (SOFC) and direct methanol fuel cells (DMFC). They can be classified on the basis of operating temperature. The low-temperature FCs, with an operating temperature range of 50–250 °C, include PEMFC, AFC and PAFC. MCFC and SOFC operate in a high temperature range of 650–1000 °C.

Some successful demonstrations have been realized in the past few years for marine applications with FCs [4–7]. Until now, most of exiting FC vessels is powered by low-temperature FCs because of their zero emissions, relatively high power density and quick start up. They are especially suitable for low power ships such as small pleasure or passenger boats, because the pure hydrogen needed for this technology can be relatively easily stored on the vessel and the energetic autonomy of the boat is not a key feature. On the other hand, high-temperature FCs, such as MCFC, SOFC, PEMFC with reformer, are more suitable for large power vessels because those FC can be fed, in this case, with more traditional fuels, such as diesel or gas. Some of feasibility, assessment and design implications are discussed in [8–11].

Although FCSs have good capability to follow the power demand during steady state operating conditions, the dynamic response to transient power demands is relatively poor. The dynamic variations of power demand will impose a serious stress on the FC membrane, thus reducing the lifetime of FCSs [3]. Besides, the FCSs employed for the vessels cannot to meet the instantaneous peak power demands, because of their higher cost compared to traditional internal combustion engine. Thus, the FCS can be hybridized with a battery bank to meet the total power demand. The degree of hybridization is the ratio of FC power to the total power requirements for a hybrid electric vehicle (HEV), and it can be determined according to the role of the energy storage system. Normally the FC should provide the minimum continuous power and the energy storage system covers the peak transient demands or acceleration performance needs. For marine applications, the primary mover of the FC must provide at least the cruising power. Otherwise it leads to dramatically increased energy requirements for the energy storage system. On the other hand, boats can provide larger space for energy storage compared to HEVs, so a battery with larger capacity can be installed on board. A hybrid FC-battery power system with a single DC/DC boost converter is proposed to meet the requirements in [12]. An energy management system (EMS) based on fuzzy logic is employed to a FC/battery/supercapacitor hybrid bus to allocate the power among different sources in [13]. Wang et al. [14] showed a comparison of different energy management strategies for HEVs.

In this paper, we will focus on the use of hybrid power configurations for a low power boat. The aim of the paper is to investigate the performance of such a hybrid system which associates a FC with a battery for the specification of low-power pleasure boat propulsion. For this case study, an efficient optimal EMS based on operation states is proposed. This EMS strategy determines the operating point of each component of the system in order to maximize the system efficiency.

The paper is organized as follows: Section 2 introduces the configuration of the hybrid power system structure for a FC boat; Section 3 presents the modeling of the hybrid power system; Section 4 proposes the EMS-based operating states; the simulations and discussion are presented in Section 5; and final conclusions are summarized in Section 6.

2. DESCRIPTIONS OF THE FC BOAT

In this paper, we used a common specification which corresponds to a real pleasure boat. These common specifications correspond to the FC boat of “Alsterwasser” and can be found in [15]. This boat has been developed by the Zero Emission Ships (Zemships) project, funded by the EU [6]. The aim of the project, which begun at the end of 2006 and ended in April 2010, was to provide technical solutions that are suitable for surface vessels. The FC boat was launched in August 2008 in Hamburg. The boat is 25.5 m long and 5.20 m wide with a carrying capacity of 100 passengers. The boat is equipped with two PEMFC of 80 kW and buffer battery of 560 V/360 A·h. The boat is fuelled by a total of 50 kg of hydrogen, which is stored onboard at a pressure of 350 bar. The parameters of the boat are shown in Table 1.

Table 1. Parameter of the fuel cell (FC) boat (data extracted from [6]). PEMFC: proton exchange membrane fuel cell.

Parameters	Value	Unit
Motor peak power	120	kW
PEMFC rated power	80	kW
PEMFC voltage	140–260	V
PEMFC current	280–520	A
Battery (lead gel)	560 V/360	A·h
Displacement of water	72	t
Length	25.56	m
Width	5.20	m
Passenger capacity	100	-
Maximum speed	15	km/h

Figure 1a shows the configuration of the hybrid power system for the FC boat. The FCS is connected to the DC bus by a boost converter and battery is connected to the DC bus directly. Figure 1b shows the typical power driving cycle of the boat propulsion motor for docking and sailing [15]. Analyzing this driving cycle the main characteristics of the required power system can be evaluated:

- The propulsion peak power is around 110 kW;
- The acceleration time is around 32 s (163–195 s);
- The cruising power is around 40 kW.

The average power is around 41 kW, without considering auxiliary power during the power cycle.

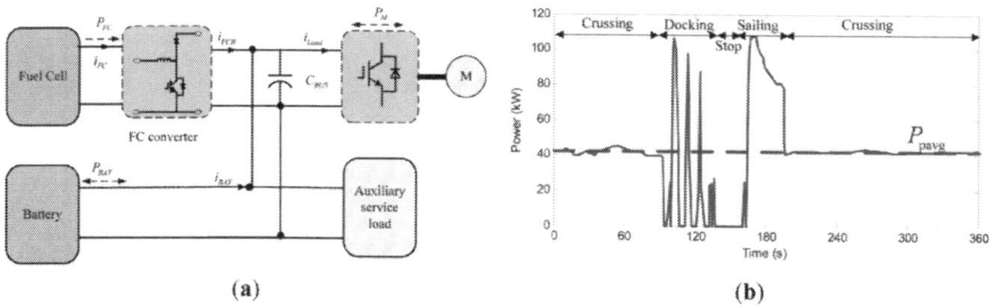

Figure 1. (**a**) Configuration of the hybrid power system for the FC boat; and (**b**) typical load characteristics for the propulsion motor [15].

3. MODELING OF THE FC/BATTERY HYBRID ENERGY SYSTEM

3.1. PEMFC

The FCS is an electrochemical device that converts chemical energy directly into electrical energy, which involves a set of physical and chemical reactions. Figure 2 shows the schematic of the FC. The FC is composed of anode, electrolyte and cathode. The electrolyte of PEMFC is a solid phase polymer membrane, which can conduct protons or H^+ ions thanks to its excellent selective permeability.

The hydrogen fuel is fed continuously to anode electrode. Oxygen is simultaneously fed to cathode. Protons and ions are produced with an oxidation reaction shown in Equation (1). The positive ions flow from anode to cathode by the electrolyte exchange membrane which does not conduct electrons. If an external circuit is connected electrons can move through this circuit creating a current flow. All the positive or negative ions from anode to cathode combine with oxygen to produce water shown in Equation (2). The overall chemical reaction is shown by Equation (3).

Oxidation half reaction:

$$2H_2 \rightarrow 4H^+ + 4e^- \qquad (1)$$

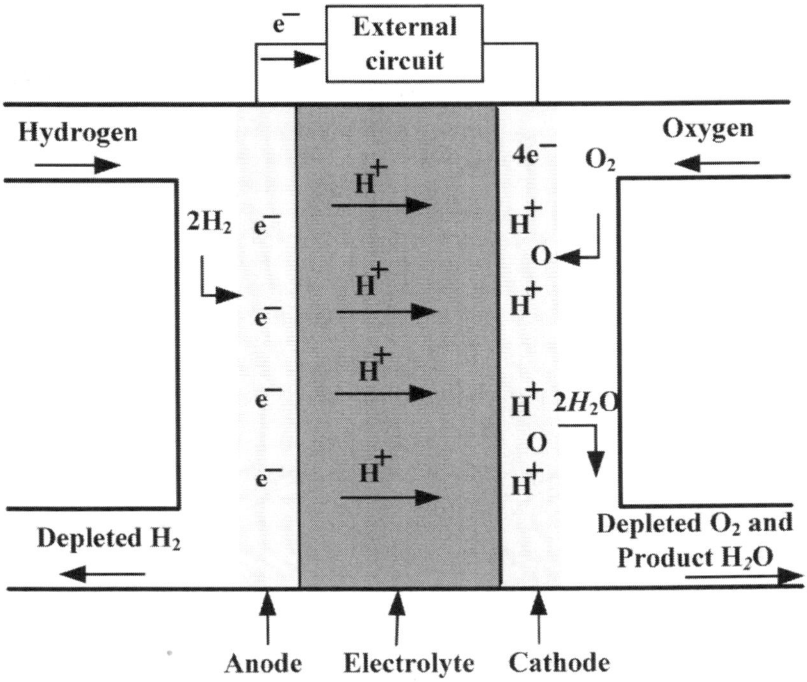

Figure 2. Schematic of an individual PEMFC.

Reduction half reaction:

$$O_2 + 4H^+ + 4e^- \rightarrow 2H_2O \tag{2}$$

Cell reaction:

$$2H_2 + O_2 \rightarrow 2H_2O \tag{3}$$

In this paper, the following assumptions are considered to obtain a simplified FC model:

- The stack is fed with hydrogen and air and the gases are considered to be ideal;
- The pressure drops across flow channels are considered to be negligible;

- The stack operates at a constant temperature of 65 °C;
- The humidity inside the stack is kept constant;
- The stack is assumed not to work in the mass transport operating conditions.

The ideal open circuit voltage for a FC can be obtained by the Nernst equation, which is influenced by the partial pressures of reactants and products, temperature, reactant concentrations. Then, the output voltage of an FC is function of the Nernst's voltage, activation voltage, ohmic voltage drop, and concentration voltage [17,18]. Thus, a FC stack can be modeled by a controlled voltage source in series with a constant resistance. The FC voltage value can be defined as follows:

$$V_{fc} = E_{OC} - NA\ln(\frac{i_{fc}}{i_0}) - i_{fc} r_{fcin} \tag{4}$$

where V_{fc} is output voltage (V); E_{OC} is open circuit voltage (V); N is number of the cells; A is the Tafel slope; i_{fc} is output current (A); i_0 is exchange current (A); and r_{fcin} is internal resistance (Ω). The open circuit voltage can be defined as follows:

$$E_{OC} = K_C E_n \tag{5}$$

$$E_n = 1.229 + (T - 298)\frac{-44.43}{2F} + \frac{RT}{2F}\ln(P_{H_2} P_{O_2}^{1/2}), \quad T \le 100°C \tag{6}$$

where K_C is voltage constant, which can be determined by internal current and the Tafel slope under nominal operation condition and the value is less than or equals to one; R is gas constant (8.3145 J/mol K); T is temperature of operation (Kelvin); F is Faraday constant (96,485 A s/mol); P_{H2} is partial pressure of H_2 (atm); and P_{O2} is partial pressure of O_2 (atm). The partial pressures of gases can be calculated as follows:

$$P_{H_2} = (1 - U_{f_{H_2}})x\% P_{fuel} \tag{7}$$

$$P_{O_2} = (1 - U_{f_{O_2}})y\% P_{air} \tag{8}$$

where U_{fH2} is hydrogen conversion rate; U_{fO2} is oxygen conversion rate; x% is the percentage of hydrogen in the fuel; y% is percentage of oxygen in the oxidant; and P_{fuel} and P_{air} are supply pressure of hydrogen and air, respectively. The rates of conversion of hydrogen and oxygen are determined as follows:

$$U_{f_{H_2}} = \frac{60000RTNi_{fc}}{2FP_{fuel}V_{lpmf}x\%} \tag{9}$$

$$U_{f_{O_2}} = \frac{60000RTNi_{fc}}{4FP_{air}V_{lpma}y\%} \tag{10}$$

where V_{lpmf} and V_{lpma} are the flow rates of hydrogen and air in L/min.

The studied FC is a PEMFC with a rated power of 80 kW. This FCS's parameters are summarized in Table 2.

Table 2. Parameters of PEMFC.

Parameters	Value	Unit
Nominal stack power	80	kW
Nominal stack efficiency	54.5%	-
FC resistance	0.070	Ω
Nernst voltage of a cell	1.1125	V
Exchange current	0.50886	A
Exchange coefficient	0.77139	-
Fuel composition	99.99%	-
Oxidant composition	21%	-
Operating temperature	65	°C

The characteristics of this studied device (V-I and P-I characteristic curves) are shown in Figure 3.

In a FCS, auxiliary power is often necessary because of the auxiliary components, such as compressor and cooling system. The auxiliary power requirement is estimated to be up to 30% of the FC stack power. The compressor consumes the main part (as much as 93.5%) of this auxiliary system power [19]. The typical power curves of the FC stack and FCS are shown in Figure 4a only considering the compressor power consumption. It can be found that the output power is approximately proportional to stack current. The efficiency of the FCS strongly depends on the output power, where the auxiliary power consumption is taken into account in this work, shown in Figure 4b. From Figures 3 and 4, it can be found that the approximate stack efficiency of the FC stack is 50% at the power of 98 kW, but the FCS efficiency is only 42%.

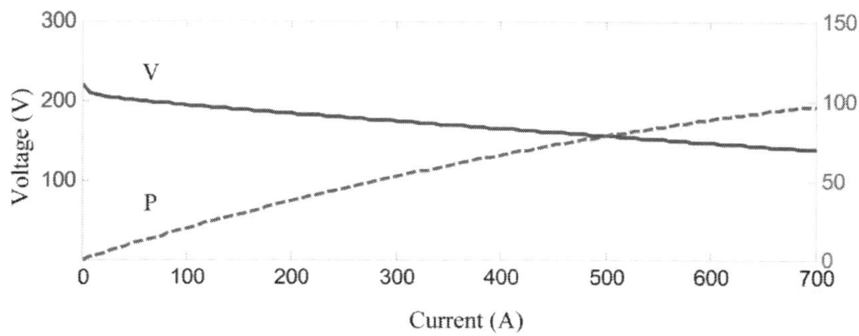

Figure 3. Typical characteristics of voltage, power vs. current.

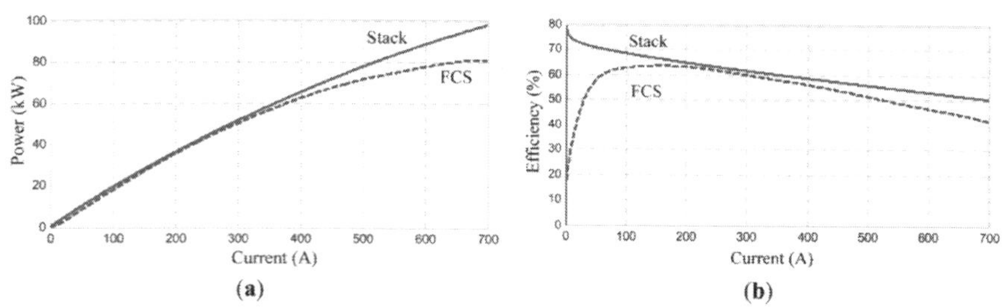

Figure 4. (**a**) Relationship between stack current and power; and (**b**) relationship between stack current and efficiency. FCS: FC system.

The FC voltage will drop when the output current arises. Therefore it should not to be connected to load directly, and a DC/DC converter is needed to control the output voltage.

3.2. Battery Bank

In our study, a lithium-ion battery has been chosen. This kind of battery is an attractive candidate for electric vehicle (EV) applications (bus, truck and boat) due to its high energy density, long cycle life and relatively low self-discharge rate, compared to traditional lead acid batteries. Therefore, in our case, a high energy polymer Li-ion battery bank is used as energy storage device. The battery is modeled classically as a controlled voltage source in series with a resistance as shown in Figure 5.

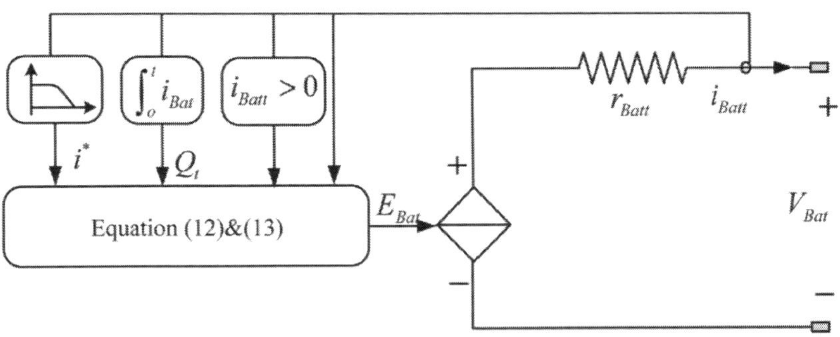

Figure 5. Model of the lithium-ion battery.

The battery output voltage (V_{Bat}) is derived as follows [19,20]:

$$V_{\mathrm{Bat}} = E_{\mathrm{Batt}} - r_{\mathrm{Bat}} i_{\mathrm{Batt}} \tag{11}$$

where E_{Batt} is no load voltage (V); i_{Batt} is battery current (A); and r_{Bat} is the internal resistance (Ω). The internal resistance is assumed constant for simplicity reasons, although it normally changes according charging and discharging conditions or value of battery current.

The open circuit battery voltage which depends on the battery current, state of charge (SOC) and hysteresis phenomenon of the battery with the temperature ignored is defined as follows. The behavioral model of this battery comprises two parts: a discharge mode model and a charge mode model which are used considering the operating conditions (testing on the value of i*).

The value of battery voltage in the discharge mode (i* > 0) is given by:

$$E_{\text{Batdis}} = E_0 - \text{K}\frac{Q}{Q-Q_t}i^* - K\frac{Q}{Q-Q_t}Q_t + A\exp(-BQ_t) \tag{12}$$

The voltage value in charge mode (i* < 0) is given by Equation (13):

$$E_{\text{Batcha}} = E_0 - \text{K}\frac{Q}{0.1Q+Q_t}i^* - K\frac{Q}{Q-Q_t}Q_t + A\exp(-BQ_t) \tag{13}$$

In Equations (12) and (13), E_{Batdis} and E_{Batcha} are the open circuit voltage (V); E_0 is constant voltage (V); K is polarization constant (V/A h); i* is low frequency current dynamics (A); i_{Batt} is battery current (A); Q_t is actual battery charge (A h); Q is Maximum battery capacity (A h); A is exponential voltage (V); and B is exponential capacity (A h).

The SOC of battery is denoted by:

$$Q(t) = Q(0) - \int_0^t i_{\text{Batt}}\,dt \tag{14}$$

C rate of battery is used to describe its discharging or charging. The C rate is defined as:

$$C\text{ rate} = \frac{i_{\text{Bat}}}{Q/h} \tag{15}$$

The studied 576 V/180 A·h battery bank is composed of three packs in parallel, and every pack is made up of 12 × 48 V/60 A·h modules in series. The parameters are listed in Table 3.

Figure 6 shows the comparison between the experimental and simulated discharge characteristics. From the figure, it can be seen that the battery voltage error is less than 5% when the battery SOC is more than 10%.

Table 3. Parameters of battery (data extracted from [21]). DOD: depth of discharge.

Type	48 V/60 A h	576 V/180 A h
Capacity	60 A h	180 A h
Output voltage	36–55 V	430–660 V
Continuous current (0.5C)	30 A	90 A
Energy	2.6 kW h	94.5 kW h
Life cycle (0.3C, 80% DOD)	1500	1500
Maximum discharging current (2C in 10 s)	60 A	360 A

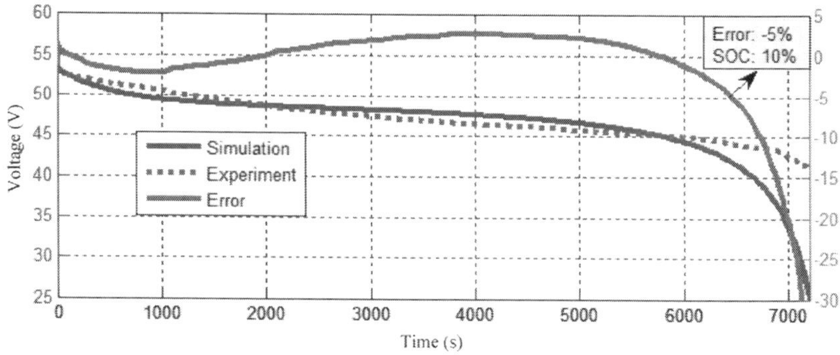

Figure 6. Comparison between experiment and simulation discharge characteristics.

3.3. DC/DC Converters

One DC/DC converter is used in this hybrid power system. A boost DC/DC converter is associated with FC and allows control the energy exchange between FC and DC bus. This converter shown in Figure 7a is composed of an inductor L, diode and a switch S.

In this study, this converter is modeled by a classical average state model with ideal components shown in Figure 7b. This basic model comprises a voltage controlled voltage source and a current controlled current source. The values of v_i and i_2 can be simply identified by d, which is the duty ratio of switch S/S_1.

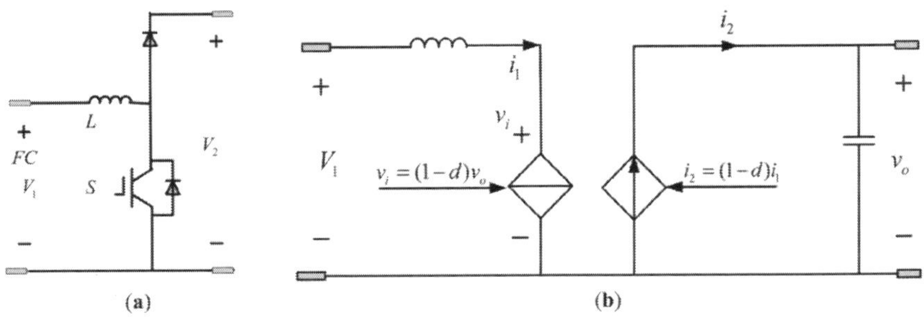

Figure 7. (**a**) Boost converter; and (**b**) average state model for DC/DC converter.

3.4. Auxiliary Services

In this paper, the auxiliary load (light and pump, etc.) is assumed constant (5 kW) and modeled for simplicity as a controlled DC current source connected directly to the DC bus. The current is a function of the power and the DC bus voltage. Then considering the driving cycle of the boat the average power is 46 kW.

4. EMS

There are two power sources in the hybrid power system. The energy management strategy for different power sources has great effect in influencing the fuel economy, dynamic performance and service life of power sources [22–24]. A states-based EMS is proposed in [22], in which the battery is connected to the DC bus by a bidirectional DC/DC convertor. A simplified EMS is proposed for a passive FC/battery hybrid power system in [23]. A cascaded control loop with a decoupling strategy in the frequency domain is proposed for FC/supercapacitor hybrid power system in [24].

The proposed EMS is used to control the energy flow between the FC and the battery, which depends on the operating states considering the optimum power of FC and battery. The operating condition determination is based on the load power, the battery, and the FC states and operating points.

Figure 8 shows the relationship between FCS efficiency and output power. The maximum FCS efficiency is about 64% and the FCS can operate with an efficiency higher than 60% over a large range of powers. When the system operates at a very low output power (less than 3 kW) the FCS efficiency is quite low. To achieve good fuel economy, the FCS should be avoided to operate in poor efficiency regions. Here, the minimum FCS power P_{FCmin} is selected as 10 kW, with an efficiency of about 60%. Considering the average power of boat, the optimum power P_{FCopt} is determined as 46 kW, where the net efficiency is 61%. The maximum net power P_{FCmax} which can be delivered by the FCS is 80 kW.

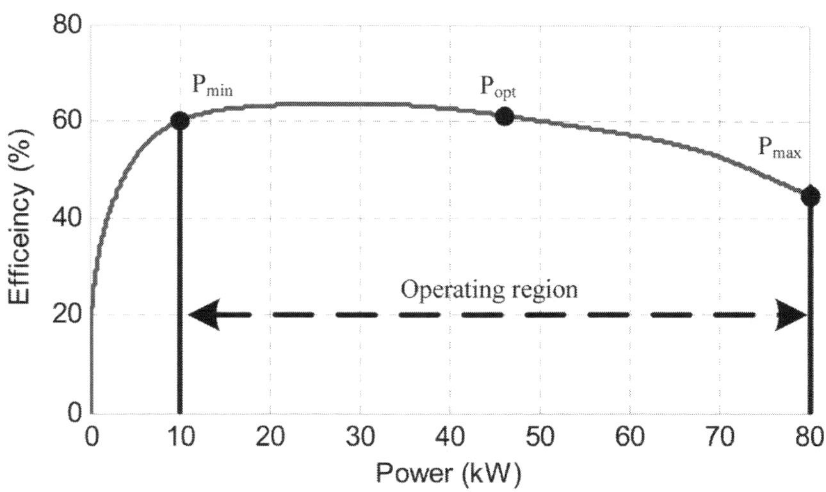

Figure 8. Relationship between FCS efficiency and power.

Because of the battery's current and voltage limits, the following battery limits have been considered: (1) charging, discharging, and optimum power values for battery ($P_{optchar}$, P_{optdis}, and P_{BATopt}), where charging power equals to discharging power with 30% capacity of battery and optimum power value is 20% capacity of the battery; and (2) battery SOC (SOC_{max} and SOC_{min}). Three levels for the battery SOC are considered: high SOC (>80%), normal SOC (50%–80%), and low SOC (<50%), which are shown in Table 4.

Table 4. Summay of the operating states. SOC: state of charge.

SOC of Battery	State	Power of load	Reference power of FC
$SOC > 80\%$	1	$P_{load} \leq P_{FCmin}$	P_{FCmin}/braking
	2	$P_{load} \leq P_{FCmin} + P_{optdis}$	P_{FCmin}
	3	$P_{Load} \leq P_{FCmax} + P_{optdis}$	$P_{FC} = P_{Load} - P_{optdis}$
	4	$P_{FCmax} + P_{optdis} < P_{Load}$	P_{FCmax}
$50\% \leq SOC \leq 80\%$	5	$P_{load} \leq P_{FCmin}$	P_{FCmin}
	6	$P_{load} \leq P_{FCopt} - P_{BATopt}$	P_{load}
	7	$P_{load} \leq P_{FCopt} + P_{BATopt}$	P_{FCopt}
	8	$P_{load} \leq P_{FCmax}$	P_{load}
$SOC < 50\%$	9	$P_{load} > P_{FCmax}$	P_{FCmax}
	10	$P_{Load} \leq P_{FCmax} - P_{optchar}$	$P_{load} + P_{optchar}$
	11	$P_{Load} > P_{FCmax} - P_{optchar}$	P_{FCmax}

The power control strategy can be described by 11 possible cases which correspond to several combinations between P_{load} and battery SOC:

- State 1: high SOC and $P_{Load} \leq P_{FCmin}$. In this case, the FC operates at its minimum power, then $P_{FCref} = P_{FCmin}$. The battery power will be $P_{BAT} = P_{FCmin} - P_{Load}$, therefore the battery SOC increases. The braking resistance will work if the battery reaches to the SOC_{max}.
- State 2: high SOC and $P_{FCmin} < P_{Load} \leq P_{FCmin} + P_{optdis}$. In this case, the FC operates also at its minimum power. The battery works at $P_{BAT} = P_{Load} - P_{FCmin}$. Therefore the battery SOC decreases.
- State 3: high SOC and $P_{FCmin} + P_{optdis} < P_{Load} \leq P_{FCmax} + P_{optdis}$. In this case, the FC power will be fluctuant according the load power at a power of $P_{FC} = P_{Load} - P_{optdis}$. The battery works at its optimum discharging power P_{optdis}, so that the battery SOC decreases.
- State 4: high SOC and $P_{FCmax} + P_{optdis} < P_{Load}$. In this case, the FC works at its maximum power P_{FCmax}. And the battery power will be $P_{BAT} = P_{Load} - P_{FCmax}$, which means that the discharging current rate will be more than 0.3C. The battery will not keep working in this mode for a long time, because the capacity of FC and battery has been designed appropriately.

- State 5: normal SOC and $P_{Load} \leq P_{FC\_min}$. This is similar to the Case 1. The FC operates at its minimum power, and the battery power will be $P_{BAT} = P_{FCmin} - P_{Load}$. Therefore the battery SOC increases slowly.
- State 6: normal SOC and $P_{FCmin} < P_{Load} \leq P_{FCopt} - P_{BATopt}$. In this case, the FC will work in strategy of load following. The FC reference power will be equal to the commanded load power. As the low response time of FC, the battery will compensate the transient power.
- State 7: normal SOC and $P_{FCopt} - P_{BATopt} < P_{Load} \leq P_{FCopt} + P_{BATopt}$. In this case, the FC will work in optimum power P_{FCopt}. The FCS power will be constant and the battery will compensate the transient power.
- State 8: normal SOC and $P_{FCopt} + P_{BATopt} < P_{Load} \leq P_{FCmax}$. This case is similar to the State 6, the FC power is regulated to follow the load power.
- State 9: normal SOC and $P_{FCmax} < P_{Load}$. This state is similar to State 4, where the FC operates at maximum power P_{FC_max}, and the battery power will be $P_{BAT} = P_{Load} - P_{FCmax}$, therefore the battery SOC decreases.
- State 10: low SOC and $P_{Load} \leq P_{FCmax} - P_{optchar}$. In this state, the FC will provide the power of load and will charge the battery. Therefore the battery SOC will increase with the optimum charging current.
- State 11: low SOC and $P_{FCmax} - P_{optchar} < P_{Load}$. In this state, the FC power will be limited to the maximum power. And the battery charging power is $P_{FCmax} - P_{Load}$ if the FC maximum power is more than the load power. Otherwise the battery will discharge to keep the balance of the power. It should be noted that the battery will have to stop working if the battery SOC decrease to the SOC_{min}.

Figure 9 shows the required power of FC depending of the different operating conditions and SOC.

Figure 10 shows the total model of the hybrid power system of the boat. The FC voltage will be fluctuating according to the load, which is why a DC/DC converter is needed to control the output power. A current control loop is used to regulate the FC output current. The FC reference current is determined by the previously exposed power management strategy according to the operating state of the hybrid power system. The proportional-integral (PI) controller determined the duty cycle of the boost convert using the error between the FC reference current and the instantaneous current through a low pass filter.

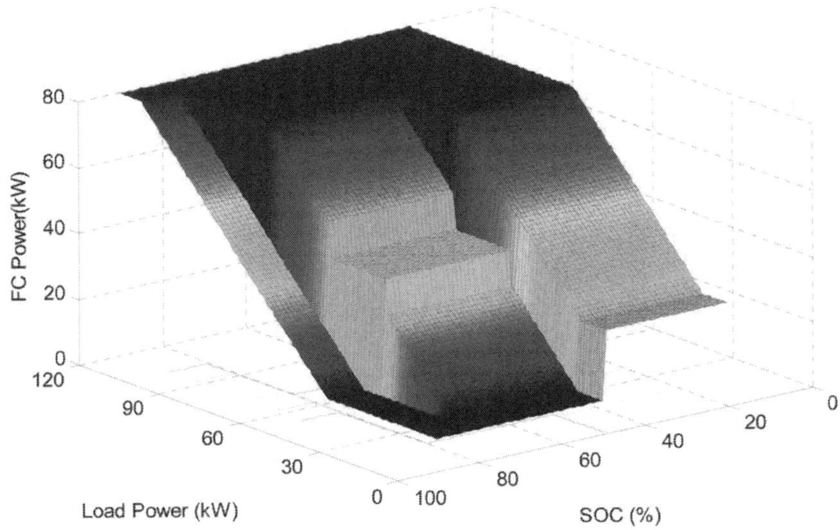

Figure 9. Power map of FC.

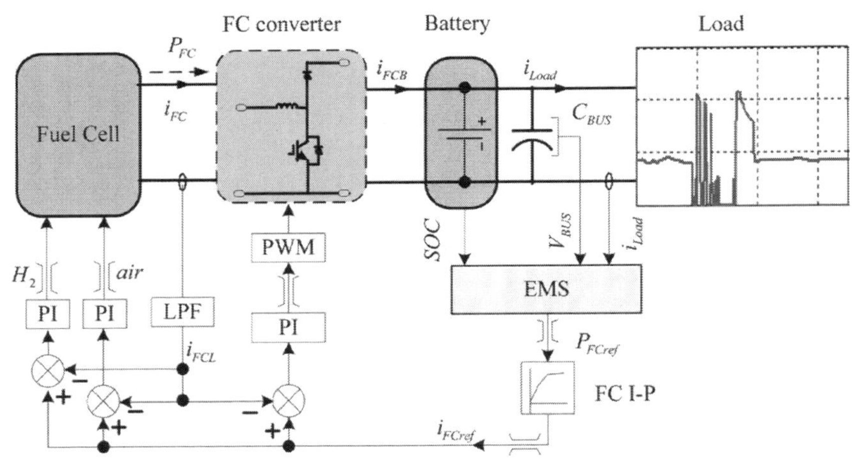

Figure 10. Model of the hybrid system. PWM: pulse width modulation; LPF: low-pass filter; PI: proportional-integral controller; and EMS: energy management system.

5. SIMULATION RESULTS

The hybrid system model is shown in Figure 10 by implementing the detailed mathematical models of the system previously described in a MATLAB/Simulink® environment. Simulation results are obtained based on the typical driving cycle of the studied passenger boat presented in Figure 1b with auxiliary load. A 300 s driving cycle is considered. It includes docking operations, stop time and cruising time. The boat begins first with a cruising period. It docks after 92 s. The duration of the docking operation is about 46 s. Then, the boat stops during approximately 24 s. Finally, it begins another cruising period and reaches cruising speed after a 35 s acceleration time. The simulation results are presented in Figures 11, 12, 13, 14, 15, 16, 17, 18, 19, 20, 21 and 22 considering different initial battery SOC: 45%, 65%, and 85%. Those three cases are designated as Case 1, Case 2 and Case 3 in the following part. The results validate the PMS of the hybrid power system under different states.

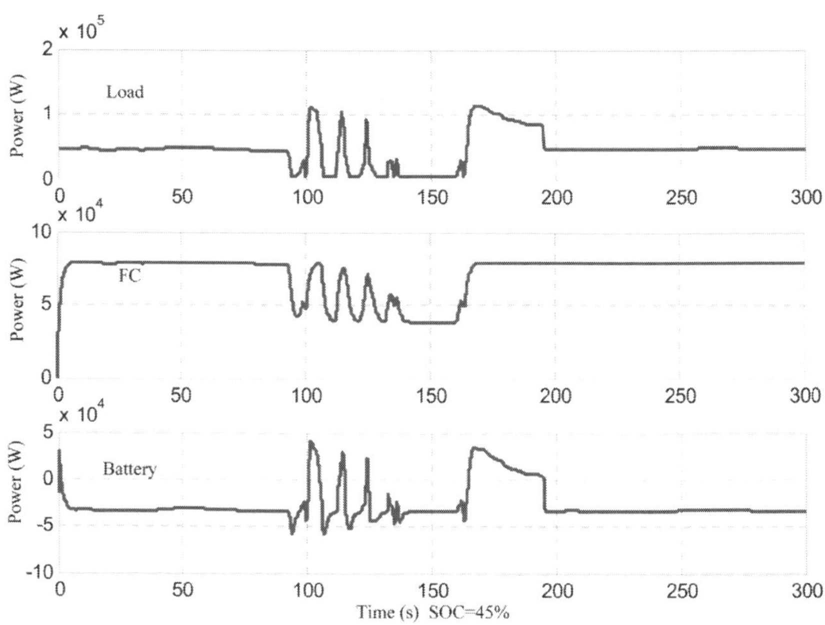

Figure 11. Power of load, FC and battery for Case 1.

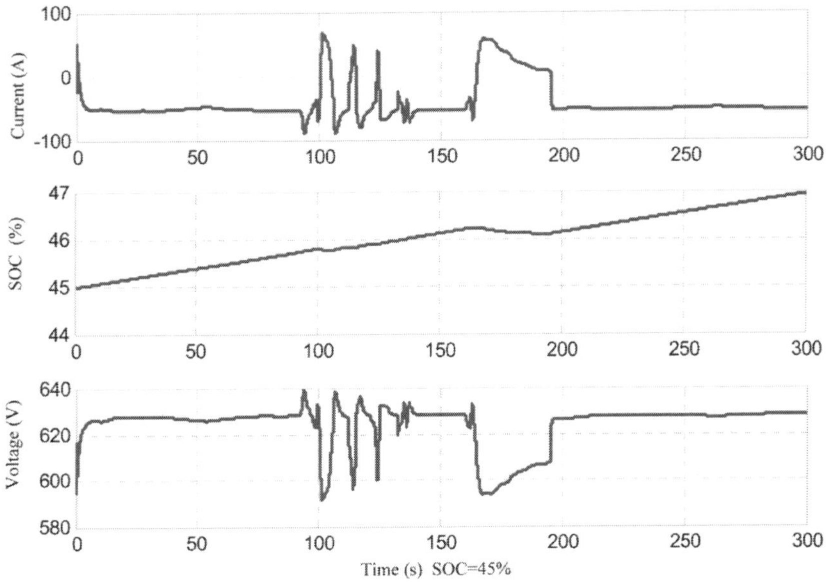

Figure 12. Battery characteristics for Case 1.

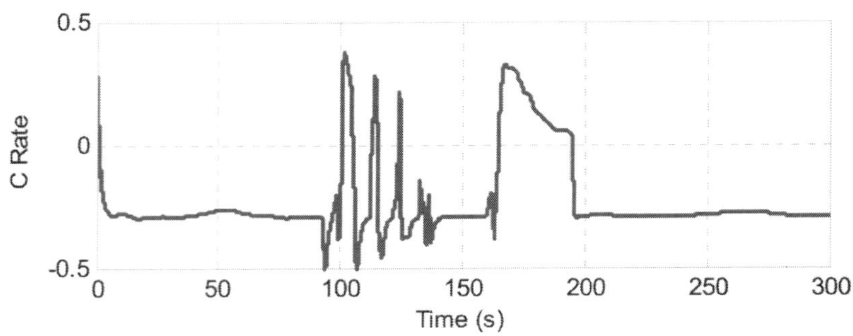

Figure 13. C rate of battery for Case 1.

In this case, the initial battery SOC is 45%, so that the battery begins to work with a low SOC. Therefore the FC will try to charge the battery to increase the SOC if the load is not too high. Figure 11 shows the power of the different components of the hybrid power system.

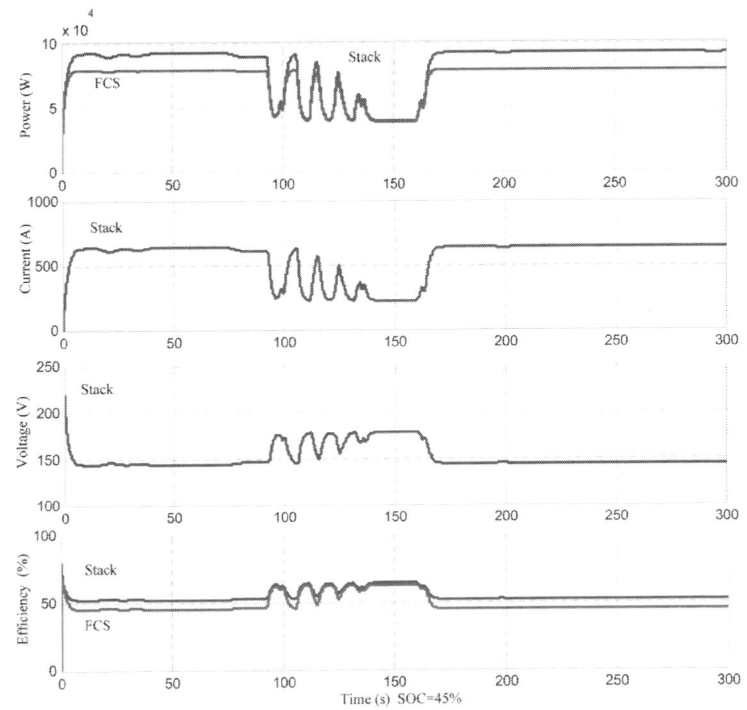

Figure 14. FC characteristics for Case 1.

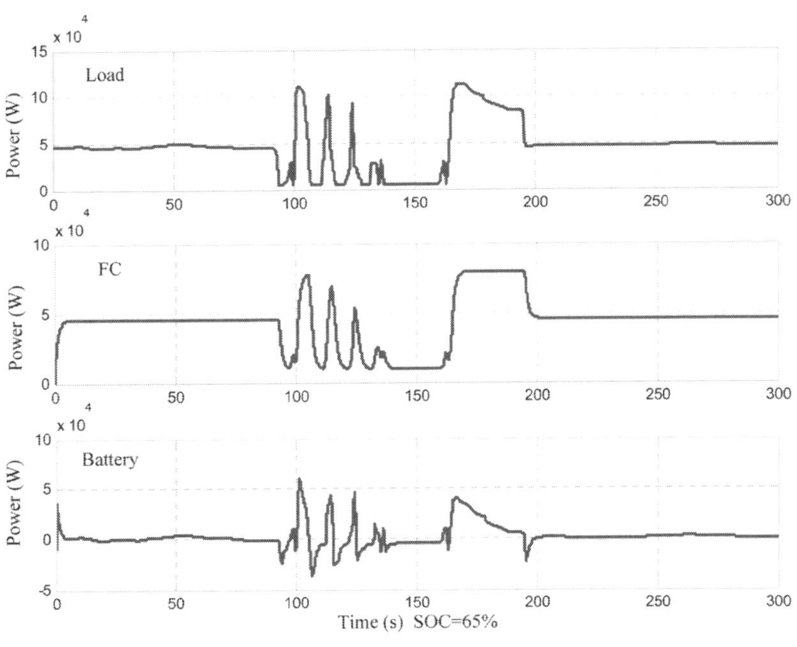

Figure 15. Power of load, FC and battery for Case 2.

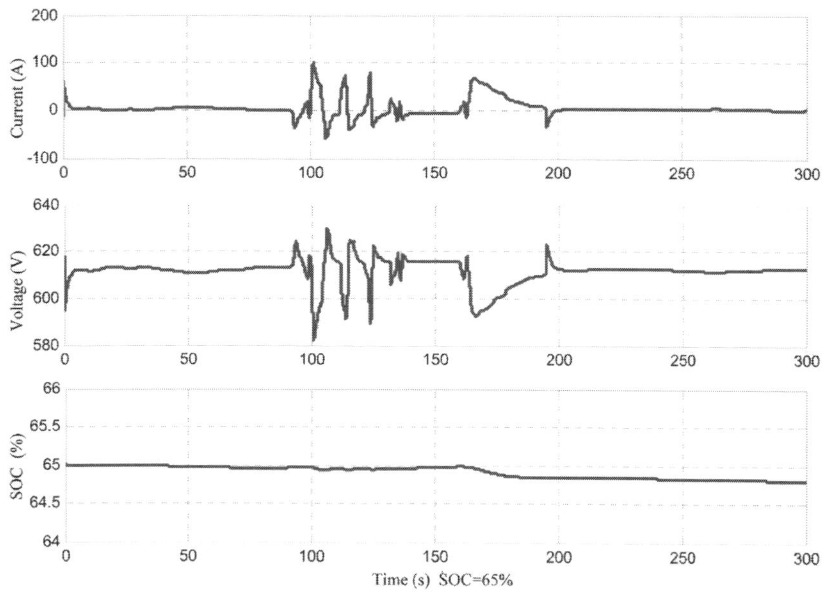

Figure 16. Battery characteristics for Case 2.

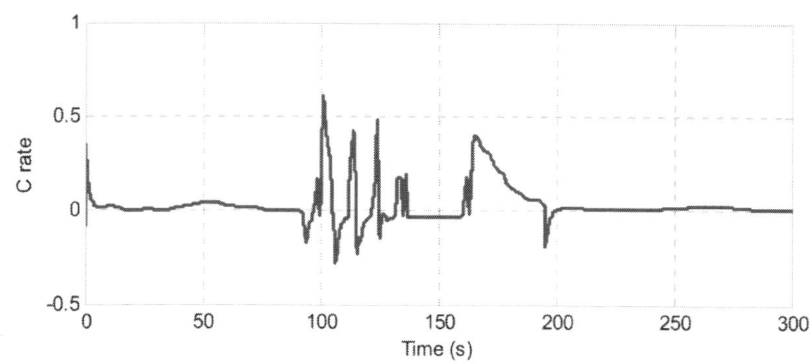

Figure 17. C rate of battery for Case 2.

5.1. Case 1: Simulations with Initial Battery SOC of 45%

Figure 12 shows the battery current, voltage and SOC. It should be noted that the positive current of battery means that the battery supply the loads and the negative one means that the battery is charged. It can be noticed that, in this case, the bus voltage fluctuates in an acceptable range of 590–640 V.

Figure 13 shows the C rate (charging/discharging rate) of the battery. In the figure, it can be found that the maximum discharging/charging rates are 0.36 and 0.5, respectively. The continuous discharging time with C rate greater than 0.3 is less than 4 s.

Figure 14 shows the FC characteristics for Case 1. The FC stack characteristics are shown in blue line. The magenta lines show the net power and efficiency. The FCS efficiency is then more than 45% during all the cycle.

As shown in the Figures 11, 12, 13 and 14, FC tries to charge the battery at the power of $P_{optchar}$ during the cruising and stop conditions. Then the battery SOC increases slowly, and it reaches to 47% after the full cycle. During docking, the FCS power is less than 50 kW for the long response time. Then, the battery bank compensates the transient load power and absorbs the extra power from the FCS. During the acceleration, FC reaches the maximum output power.

5.2. Case 2: Simulations with Initial Battery SOC of 65%

In this second case, the initial battery SOC is 65%. It means that the battery begins to work with a normal SOC. Therefore the FC will work mostly at the command of load power and sometimes at the optimum power.

Figure 15 shows the power of the different components of the hybrid power system. When the boat cruises, the FC works at the optimum power. The maximum powers of FCS and battery during docking are 76 kW and 62 kW, respectively.

Figure 16 shows the battery current, voltage and SOC for Case 2. The battery SOC is kept constant and the bus voltage also fluctuates in an acceptable range of 580–620 V.

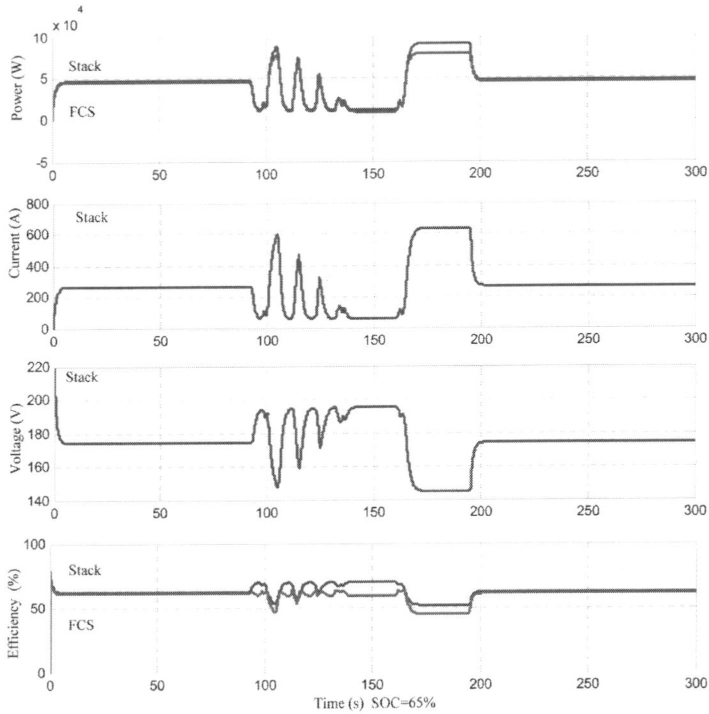

Figure 18. FC characteristics for Case 2.

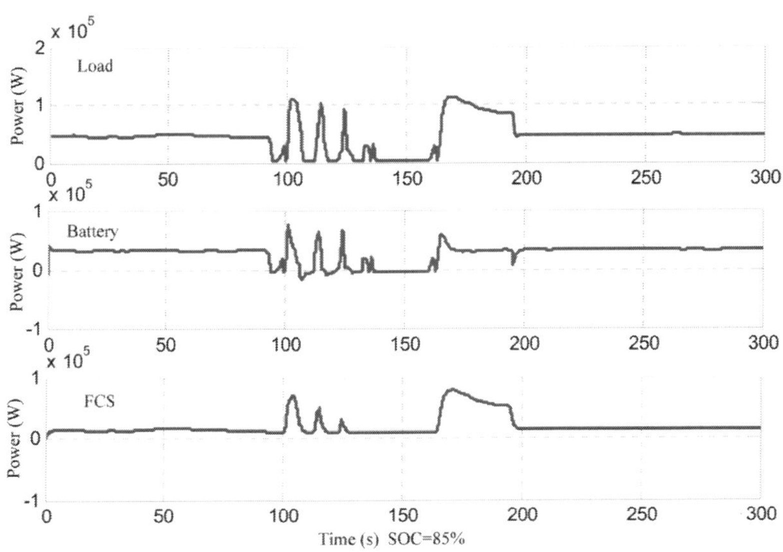

Figure 19. Power of load, FC and battery for Case 3.

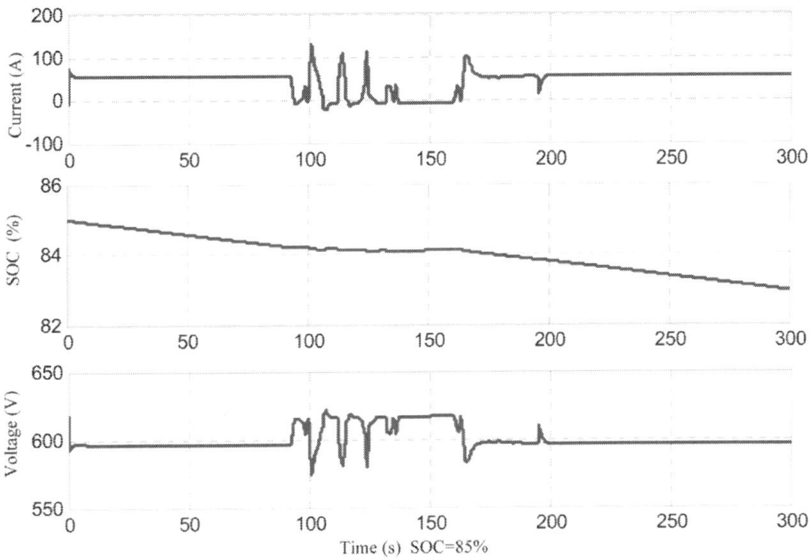

Figure 20. Battery characteristics for Case 3.

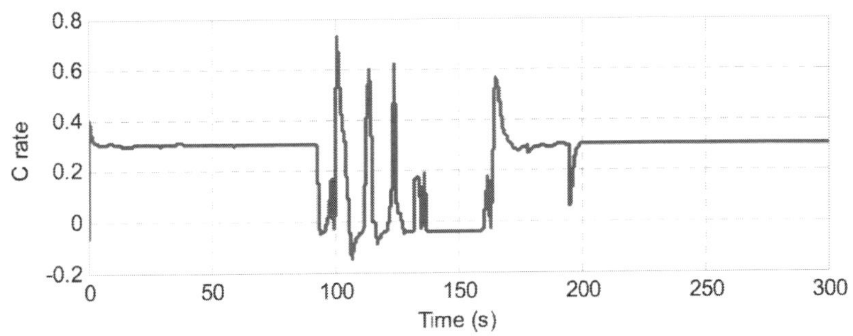

Figure 21. C rate of battery for Case 3.

Figure 17 shows the C rate of battery during the cycle. In the figure, the maximum discharging rate is 0.6 at 100 s. The charging rate is less than 0.3 during the cycle.

Figure 18 shows the FC characteristics for Case 2. The FC stack characteristics are shown in blue line. The magenta lines show the net power and efficiency. The FCS efficiency is about 61% mostly, and it is about 45% under acceleration conditions.

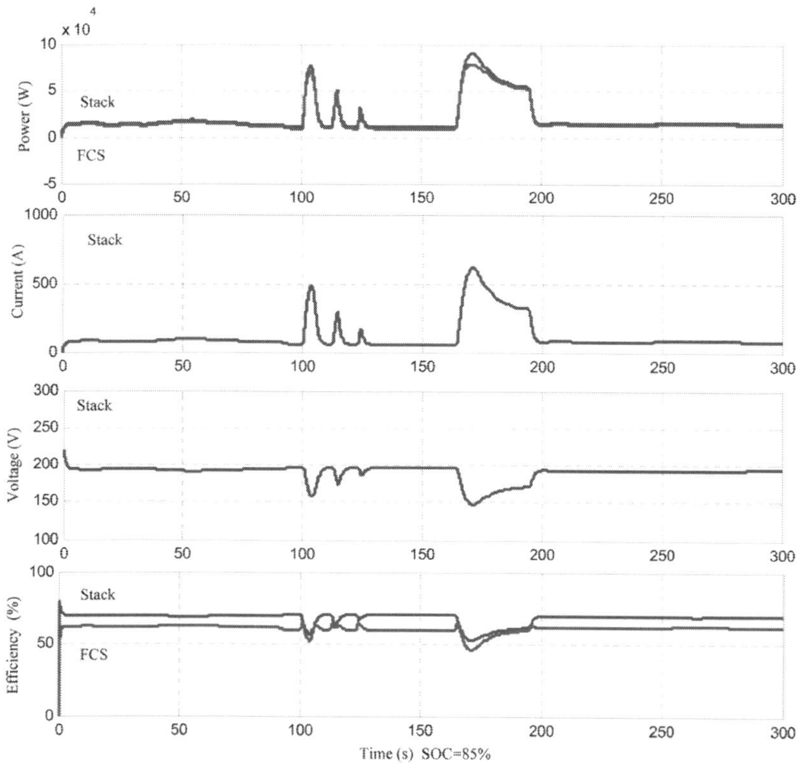

Figure 22. FC characteristics for Case 3.

5.3. Case 3: Simulations with Initial Battery SOC of 85%

In this case, the initial battery SOC is 85%, therefore, the power management strategy leads to discharge the battery to a reach a normal SOC. Figure 19 shows the power of the different components of the hybrid power system for Case 3. When the boat cruises, the battery discharges at the optimum discharging power for the high SOC. Then FCS works at a low power and its maximum power is 78 kW during acceleration.

Figure 20 shows the battery characteristics for Case 3. The battery SOC decreases slowly for continuous discharge.

Figure 21 shows the C rate of battery during the cycle. In the figure, the maximum discharging rate is 0.72 at 100 s. The total time is 5.6 s when the C rate is more than 0.5 during the cycle. The charging rate is less than 0.2 during the cycle.

Figure 22 shows the FC characteristics for Case 3. The FCS efficiency is more than 60%, mostly because of the continuous discharge of the battery.

Figure 23 shows the total hydrogen mass consumption of the FCS under Case 2. The system controlled by optimal control has a lower consumption (330 g) than traditional load commands tracking control (342 g). It means that the proposed EMS also achieved better efficiency.

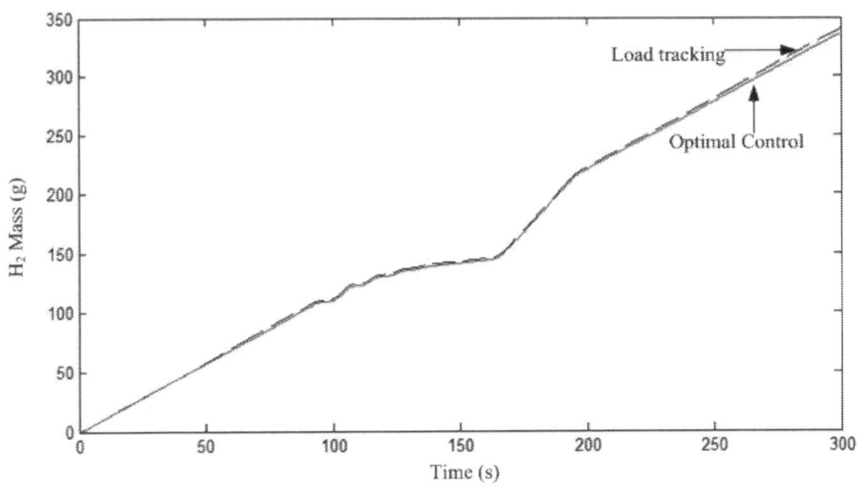

Figure 23. Hydrogen mass consumption.

6. CONCLUSIONS

There is a focus on all-electric ships powered by FCs because of their high efficiency and low emissions, and there are some R&D projects aiming to develop this technology. This paper presents a hybrid FC/battery power system for a low power boat. The hybrid system consists of PEMFC and a Li-ion battery bank. The FCS is connected to a DC bus by a boost converter, whereas the battery is connected directly to the DC bus. Mathematical models for the components of the hybrid systems are presented, which are implemented in Matlab/Simulink environment. An optimal EMS based on operation states is proposed to determine the operating point of each component of the system and optimize the system efficiency. The hybrid system behavior is then evaluated based on the real driving cycle of a low power FC boat, and the FCS works with optimized efficiency. Simulation results show that the proposed EMS can achieve higher efficiency compared with load commands tracking control, and the discharging and charging rate can be mostly controlled at less than 0.3C.

ACKNOWLEDGMENTS

This work was partly supported by the National Natural Science Foundation of China under Grant No. 51007056 and by Brittany Region (France) funds for Post-Doctoral Foreign research stay.

CONFLICTS OF INTEREST

The authors declare no conflict of interest.

REFERENCES

1. Miola, A.; Ciuffo, B.; Giovine, E.; Marra, M. Regulating Air Emissions from Ships: The State of the Art on Methodologies, Technologies and Policy Options; Publications Office of the European Union: Luxembourg, 2010. Available online: http://ec.europa.eu/dgs/jrc/downloads/jrc_reference_report_2010_11_ships_emissions.pdf (accessed on 10 September 2012).

2. Tsea, L.K.C.; Wilkinsa, S.; McGlashana, N.; Urban, B.; Martinez-Botas, R. Solid oxide fuel cell/gas turbine trigeneration system for marine applications. J. Power Sources **2011**, 196, 3149–3162.

3. EG&G Technical Services, Inc. In Fuel Cell Handbook, 7th ed.; US Department of Energy, Office of Fossil Energy, National Energy Technology Laboratory: Morgantown, WV, USA, 2004.

4. Tang, T.; Han, J.; Yao, G.; Feng, Y. Development of a PEM Fuel Cell Boat. Proceedings of the 12th International Power Electronics and Motion Control Conference (EPE-PEMC), Portoroz, Slovenia, 30 August–1 September 2006.

5. H2Yacht. Available online: http://www.h2yacht.com/ (accessed on 4 September 2012).

6. Zemships—Zero Emission Ships. One Hundred Passengers and Zero Emissions: The First Ever Passenger Vessel to Sail Propelled by Fuel Cells. Available online: http://ec.europa.eu/environment/life/ project/ Projects/index.cfm?fuseaction=home.showFile&rep=file&fil=Zemsh ips_Brochure_EN.pdf (accessed on 9 January 2013).

7. Lovers Amsterdam. CO_2Zero Canal Cruise. Available online: http://www.lovers.nl/co2zero/ (accessed on 2 December 2012).

8. Viking Lady. Available online: http://www.vikinglady.no/ (accessed on 6 March 2013).

9. Alkaner, S.; Zhou, P. A comparative study of lice cycle analysis of molten carbon fuel cells and diesel engines for marine application. J. Power Sources **2006**, 158, 188–189.

10. Bensaid, S. MCFC-based marine APU: Comparison between conventional ATR and cracking coupled with SR integrated inside the stack pressurized vessel. Int. J. Hydrog. Energy **2009**, 34, 2026–2042.

11. Wei, J.; Fang, R. Performance Prediction and Dynamic Simulation of Electric Ship Hybrid Power System. Proceedings of the IEEE Electric Ship Technologies Symposium, Philadelphia, PA, USA, 21–23 May 2007.

12. Chao, C.-H.; Shieh, J.-J. A new control strategy for hybrid fuel cell-battery power systems with improved efficiency. Int. J. Hydrog. Energy **2012**, 37, 13141–13146.

13. Gao, D.; Jin, Z.; Lu, Q. Energy management strategy based on fuzzy logic for a fuel cell hybrid bus. J. Power Sources **2008**, 185, 311–317.

14. Wang, X.; He, H.; Sun, F.; Sun, X.; Tang, H. Comparative study on different energy management strategies for plug-in hybrid electric vehicles. Energies **2013**, 6, 5656–5675.

15. Mertens, A. The Zemships Propulsion System and Beyond. Proceedings of the Zemships Conference H2Expo, Hamburg, Germany, 22–23 October 2008.

16. Han, J.; Charpentier, J.F.; Tang, T. State of the Art of Fuel Cells for Ship Applications. Proceedings of the 2012 IEEE International Symposium on Industrial Electronics, Hangzhou, Zhejiang, China, 28–31 May 2012; pp. 1456–1461.

17. Souleman, N.M.; Olivier, T.; Dessaint, L.A. A Generic Fuel Cell Model for the Simulation of Fuel Cell Vehicles. Proceedings of the IEEE Power & Energy Society General Meeting, Calgary, AB, Canada, 26–30 July 2009.

18. SimPowerSystems; MathWorks Inc.: Natick, MA, USA, 2010.

19. Boettner, D.D.; Paganelli, G.; Guezennec, Y.G.; Rizzoni, G.; Moran, M.J. Proton exchange membrane fuel cell system model for automotive vehicle simulation and control. J. Energy Resour. Technol. **2002**, 124, 20–27.

20. Tremblay, O.; Dessaint, L.-A.; Dekkiche, A.-I. A Generic Battery Model for the Dynamic Simulation of Hybrid Electric Vehicles. Proceedings of the IEEE Vehicle Power and Propulsion Conference, Arlington, VA, USA, 9–12 September 2007.

21. TOSUN Energy Inc. Battery Datasheet (48 V–60A h). Available online: http://www.tsebattery.com/china/prod_ups.asp? BigClass ID=45 (accessed on 4 May 2013).

22. Garcia, P.P.; Fernandez, L.M.; Garcia, C.A.; Jurado, F. Energy management system of fuel-cell-battery hybrid tramway. IEEE Trans. Ind. Electron. **2010**, 57, 4013–4023.

23. Chen, Y.; Lin, S.; Hong, B. Experimental Study on a passive fuel cell/battery hybrid power system. Energies **2013**, 6, 6413–6422.

24. Azib, T.; Bethoux, O.; Remy, G. An innovative control strategy of a single converter for hybrid fuel cell/supercapacitor power source. IEEE Trans. Ind. **2010**, 57, 4024–4031.

CHAPTER 5

Naval Engineering: Data Indicative of the Material State

Gary Ford[a*], Chris McMahon[a], Chris Rowley[b]

[a]University of Bristol,Bristol, UK

[b]Babcock International, Keynsham, UK

ABSTRACT

Naval vessels are complex artefacts, containing in excess of 100 integrated hard systems which are linked structurally, mechanically, electrically, hydraulically, pneumatically and electronically. The systems may be configured to provide a variety of prescribed capabilities and associated command objective. However, the hard systems will not fully integrate or function in a cohesive manner without the interaction of operators or maintainers, i.e. soft systems. The complexity of a naval system is not just the assimilation of hard and soft systems but also the consequence of numerous internal and external influences, e.g. system capability / constraints, maritime doctrine. Within the In-Service lifecycle stage, a naval vessel will cycle through 3 discrete phases, i.e. Upkeep, Regeneration and Tasking. Each cyclical phase will generate volumes of structured, unstructured, objective, subjective data. The assessment of the material state of a system invariably requires multiple data sources assimilated to provide a perspective, i.e. *Weltanschauung*. Furthermore, within each cyclical phase the influence, responsibility, requirements and individual perspective of stakeholders will vary, e.g. Customer ~ Ministry of Defence, Operated by ~ the Royal Navy, Maintained by ~ Babcock / BAE Systems. Furthermore, data may be seen to permeate and be utilised across each cyclical phase. The information sources identified during the research are analysed to assess the potential value of the information contained within which includes a consideration of the levels of abstraction and utilisation.

KEYWORDS

Maintenance; Data; Naval

1. INTRODUCTION

Royal Navy (RN) vessels may be considered to be complex systems of systems, i.e. whose system elements are themselves systems; typically these entail large scale inter- disciplinary problems with multiple, heterogeneous, distributed systems [1]. Naval vessels are "unreliable in the sense that they deteriorate with age and/or usage and ultimately fail" [2] thus necessitating constant upkeep. Upkeep will encompass not only corrective and preventive maintenance to preserve capability, but also upgrades "capable of increased functionality" to obviate challenges with respect to technology, mission, economic and threat [3,4]. Naval vessels are designed for a defined Concept of Operations (CONOPS); however, changes in mission and threat have often resulted in a revised *modus operandi* not fully considered within the original CONOPS, e.g. the change from a cold war role to counter-terrorism.

The impact of a revised *modus operandi* upon systems and by implication maintenance can be dramatic. HMS Illustrious, initially conceived as an anti-submarine carrier to operate in the North Atlantic, whilst recently deployed in the Arabian Gulf noted, "Weather conditions were good, with the outside ambient air temperature over 35° Centigrade. Unfortunately, this meant that temperatures inside the machinery spaces regularly exceeded 50° Centigrade" [5]. Combined with the high tempo of RN operations from the Arctic Circle to the Arabian Gulf the load upon systems can be considerable.

During the In-Service stage a range of data / information is generated and utilised to ascertain the material state of systems. The objective of the research and this paper has been to identify the data elements, their relevance and potential value within the maintenance domain. The information sources will be used to determine the asset condition and enable improved decision making with respect to maintenance.

The research method comprised of semi-structured recorded interviews with stakeholders within each cyclical phase (process) of the In-Service lifecycle stage. The stakeholders have included Babcock surface ship and submarine Project Managers, a Class Output Management (COM) Project team, numerous RN personnel within the Force Generation Authority, Maritime Capability, Trials and Assessment, Flag Officer Sea Training, operational vessels, an equipment supplier and Defence Equipment and Support (DE&S) spares and finance managers. The interviews were focused upon identifying each stakeholder's role within the In-Service stage and in particular the stakeholders particular inputs, outputs, mechanisms (enablers) and controls within each cyclical phase / process (Figure 1).

Although the research has been data centric and intended to identify the inputs and outputs; it should be noted they function as a consequence of controls and mechanisms impacting upon the process, e.g. policies, procedures, suitably qualified and experienced personnel, test equipment, etc. The research has included a review of various information sources utilised and collected within the In-Service stage.

The paper will provide a background overview of the lifecycle including a consideration of the value of information. The cyclical In-Service processes are described including detailing many of the key data elements. Each of the information sources identified in the research are identified and analysed with respect to their potential value. This is followed by a review of information abstraction within the maintenance domain with a final conclusion.

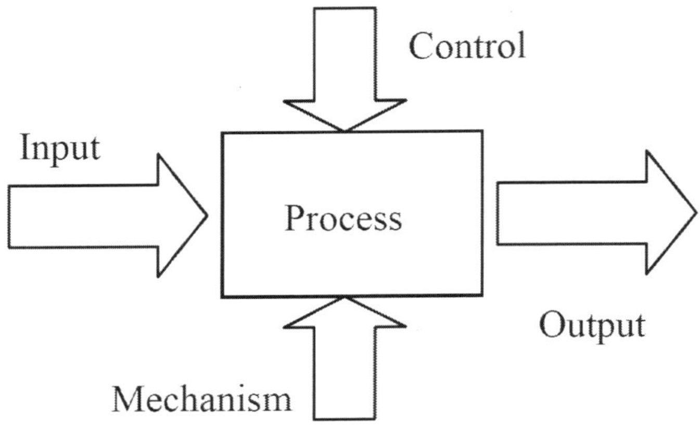

Figure 1: Process Function Model

2. BACKGROUND

The term In-Service relates to a stage within the Ministry of Defence Acquisitions Operating Framework [6]. The development and acquisition of equipment capability follows a lifecycle known as CADMID, similar in principle to ISO 15288, i.e. Systems Engineering - System Life Cycle Processes [1]. The 6 CADMID stages are, Concept, Assessment, Demonstration, Manufacture, In-Service, Disposal [7].

Within the In-Service stage, a cyclical process may be seen to exist, whereby a vessel will cycle between Upkeep, Regeneration and Tasking (Figure 2).

The In-Service phases are,

- Upkeep: deep maintenance that may include docking
- Regeneration: the activation of a vessel into a vessel capable of undertaking operational tasks.
- Tasking: undertaking or available for operational tasking, e.g. humanitarian relief, anti-piracy

- A vessel will repeat this sequence, undertaking regular Upkeep maintenance and life extension programmes between operational Tasking.

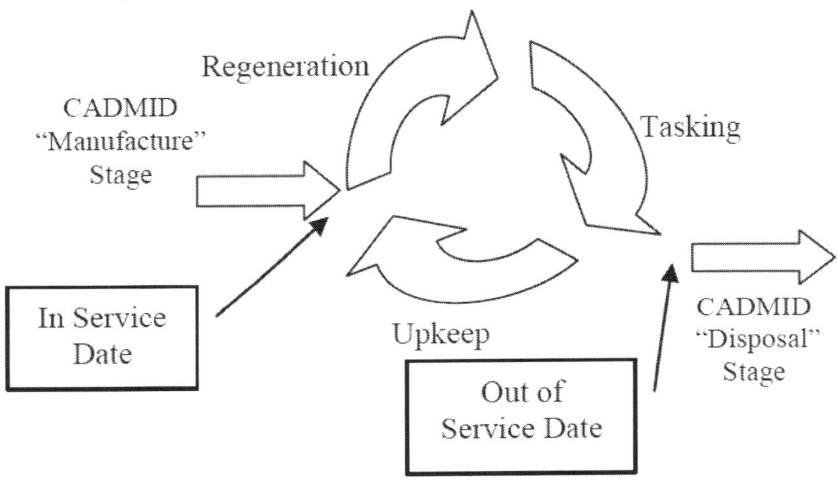

Figure 2: Cyclical In-Service Phases

Organisations are often described as, "data rich but information poor" [8,9,10,11]; this characteristic may also be ascribed to naval vessels and their support. Indeed, large volumes of data are generated throughout the life of a naval vessel, the primary intention being to improve decision making. Information may be collected as a consequence of external requirements, e.g. regulatory, in which case the value of the information and subsequent decision making exists within the external domain, e.g. Office of Nuclear Regulation. It should be noted, "the contribution that information makes to decision-making has to be made explicit" [12]. Within the Upkeep phase, determining the economic benefit resulting from better decision making is potentially achievable by undertaking a cost benefit analysis (Figure 3).

Within a naval domain, formulating a cost benefit may not always be possible, e.g. if the vessel is undertaking humanitarian relief there is no cost benefit if the main armament is not available. Hence, the *value* of information relates to the facilitation of a deterministic evaluation, e.g.

"to achieve higher order organisational objectives and generate capabilities that create sustainable advantage" [14]. Furthermore, a number of the information sources are not created, maintained or owned by the maintenance organisation, e.g. Babcock. Indeed, security may also be an issue given the premise of information sharing within a military environment is often determined by "the need to know" [15]. Assessing the economic benefit during Tasking or Regeneration may not be practicable given the objective is not financial but the application of doctrine, i.e. "The ability to project power at sea and from the sea to influence the behaviour of people or the course of events" [16].

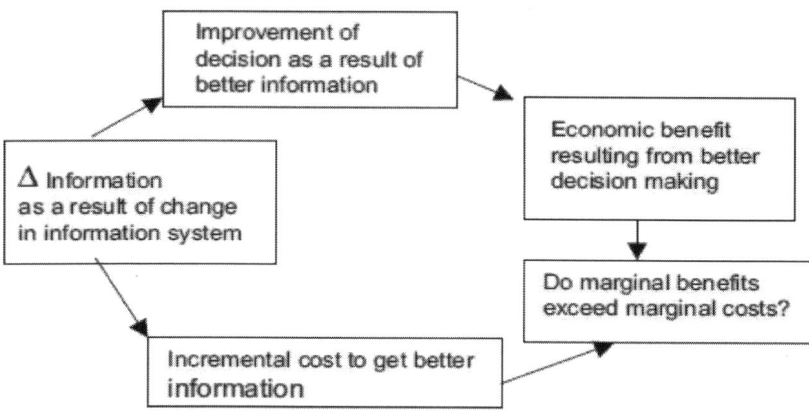

Figure 3: Benefit Chain Concept [13]

Moody and Walsh [17] extended the concept by measuring the value of information as an asset and formulated "seven laws of information" (Table 1), i.e.

The laws require no cost benefit analysis and hence may be applied within a naval domain; thus allowing a structured assessment of an information source. Analysis of the results would provide an indication of the potential value of the information source. Given the operational role / capability and security dimension of the domain, the validity of "Law 1" may be constrained, i.e. shareable with those who need to know.

Information will exhibit characteristics / properties similar to physical artefacts. A hard system will have characteristics, e.g. shape, dimension, material, and properties, e.g. function, durability, cost, [18]. Information generated or utilised within a system will also exhibit characteristics / properties, e.g. latency, qualitative, quantitative, objective [19] the properties may be beneficial, detrimental and potentially emergent.

Table 1. The Seven Laws of Information

Law	Definition
1	Information Is (Infinitely) Shareable
2	The Value of Information Increases with Use
3	Information is Perishable
4	The Value of Information Increases With Accuracy
5	The Value of Information Increases When Combined With Other Information
6	More Is Not Necessarily Better
7	Information is not Depletable

3. IN-SERVICE MATERIAL STATE DATA

Knowledge of the material state is essential in maintaining the availability and capability of naval vessels. This section will describe the objective of each cyclical phase and a brief overview of the process and information sources.

3.1 Upkeep

The objective of the Upkeep phase is the execution of preventive and corrective maintenance in addition to enhancing / extending a vessel's operational capability, e.g. performing Alterations and Additions (A's and A's). Due to the size and complexity of the task, numerous activities are initiated prior to the start of Upkeep. Surveying is undertaken during Tasking and performed by the RN / DE&S, e.g. Pre-Upkeep Material Assessment. Increasingly surveys are undertaken by the maintenance organisation with assistance from the Original Equipment Manufacturer (OEM) and form the material state datum. The resultant work package is a composite of numerous sources including the Unit Maintenance Management System (UMMS) detailing preventive maintenance and defects, survey reports (including statutory inspections), Work Requests generated by the ship staff, Operational Defects (OpDef's), A's and A's. Each of the sources provide a perspective to the material state. A key activity during Upkeep is the formation of objective information from a subjective physical environment, to readily enable decision making, e.g. if "valve leaking" then "replace". A key output from the process is the Material State Portfolio; the report details the work undertaken and records the condition of the vessel upon completion of Upkeep. The report is a mix of objective data, e.g. diesel engine hours run, and subjective assessment, e.g. "Structure: [...] the hull was assessed to be in a satisfactory condition following repair" [20]. The report also provides a datum for future assessment of hull and system degradation / usage. The majority of outputs are primarily objective, e.g. test specifications and results, certification, quality documentation; the intention being to document the maintenance package.

3.2 Regeneration

The objective is the regeneration of a vessel that is Safe to sail □ Safe to fight □ Safe to deploy. The process is initiated by a Force Generation Order (FGO) from Naval Command HQ to the Force Generation Authority (FGA). The FGA will formulate a specific user requirement based upon a vessels planned role. Following handover from the maintenance organisation, e.g. Babcock, an exhaustive series of trials to verify and validate the maintenance package and the capability of hard and soft systems, e.g. Harbour Acceptance Trials, Sea Acceptance Trials, Directed Continuation Training. The regeneration process will culminate with a FINSIG (Final Signal) and a "Ready for Ops" signal. The FINSIG contains an objective overall assessment of the vessels capability, e.g. Satisfactory, together with a detailed subjective appraisal of areas of concern and OpDef's, e.g. "The Ship Is In Reasonable Condition With 2 Cat B And 23 Cat C ME / WE OpDefs Outstanding" [21] The "Ready for Ops" signal will utilise the standard RN reporting structure, i.e. Manpower, Equipment, Training, Sustainability to assess the vessel with respect to its future role detailed in the FGO.

3.3 Tasking

The objective of the Tasking process is to undertake a range of operational tasks as directed by naval command in support of British Maritime Doctrine [16], e.g. disaster / humanitarian relief, anti-piracy patrols. The Fleet Operations Maintenance Officer (FOMO) within naval command provides engineering support for Tasking vessels. A key input to the Tasking phase and support of a vessel is the FINSIG and "Ready for Ops" signal; these reports highlight risks with respect to capability. Two distinct information domains detailing and assessing the material state of a vessel may be seen to exist, i.e.

- Onboard: generated and utilised solely within thevessel
- Onshore: generated, analysed and (potentially) aggregated onboard before transmission to onshore authorities, e.g. FOMO, for subsequent analysis, execution and provision of assistance

The onboard information may be considered tactical, i.e. single vessel, local, immediate, single stakeholder, autonomous operation, whereas onshore has a strategic focus, i.e. aggregation of multiple vessels / systems, remote, latent, long term planning, multiple stakeholders. The Common Information relates to the interface and interaction of onboard and onshore information (Figure 5).

Within the context of Tasking, the Common Information will typically comprise Fleet Operational Capability (FLOC) reports, OpDef's and Fuel and Lubrication usage reports.

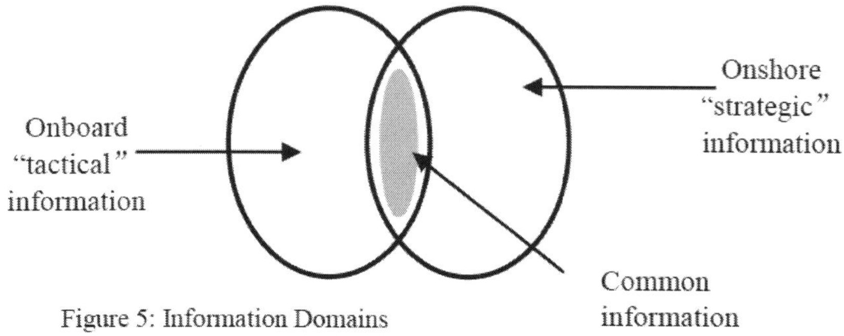

Figure 5: Information Domains

The onshore information is initially analysed by FOMO to assess and advise upon the operational capability of a vessel in relation to the perceived material state. FOMO will also coordinate and expedite support from other stakeholders, e.g. OEM, COM.

The onboard tactical information will encompass a range of engineering logs, corrective and preventive maintenance schedules and reports, materiel usage, condition based data, e.g. pressure, temperature, flow rate. The information may be considered tactical given the singular objective of sustaining the capability of the vessel by timely and effective maintenance of systems.

4. DATA ANALYSIS

Table 2 details the information sources identified as a consequence of the research, combined with the typical data operations within each cyclical phase. The data operations are, Create (*C*): insert a record, Read (*R*): select a record, Update (*U*): modify a record, Delete (*D*): remove a record.

Table 2: Information Sources and Typical Operation in Each Cyclical Phase

Information Source	Upkeep	Regen	Tasking
Alteration's and Addition's (A's & A's)	*CRUD*		
Certificates, e.g. Naval Ship Safety Cert.	*R*	*CRUD*	*CRUD*
Dynamic Machinery Trial	*R*		*CRUD*
End of Project Report	*CRUD*		
Final Signal (FINSIG)		*CRUD*	
Financial reports	*CRUD*		
Fleet Operational Capability		*CRUD*	*CRUD*
Force Generation Order		*R*	
Fresh Water Coolant Analysis	*R*	*CRUD*	*CRUD*
Fuel Analysis		*CRUD*	*CRUD*
Fuel and Lubrication Report (FLUBCON)		*CRUD*	*CRUD*
Harbour Acceptance Trials		*CRUD*	*R*
Lloyds Survey (CD Live)	*CRUD*		
Logistics Log		*CRUD*	*CRUD*
Material State Portfolio (MSP)	*CRUD*	*R*	*R*
Material Usage	*CRUD*	*CRUD*	*CRUD*
ME Defect Log		*CRUD*	*CRUD*
Night Rounds Report		*CRUD*	*CRUD*
Non-Destructive Testing, e.g. X-Ray, Ultrasound	*CRUD*	*R*	*R*
Operational Capability Confidence Check		*CRUD*	*CRUD*
Operational Defects	*CRUD*	*CRUD*	*CRUD*
Pre-Upkeep Material Assessment	*RUD*		*CRU*
Pre-Upkeep Weapon System Assessment	*RUD*		*CRU*

Ready-For-Ops Signal		*CRUD*	*R*
Report of Machinery Trials in Ships		*CRUD*	
Risk Register	*CRUD*		
S2022's Defects	*R*	*CRUD*	*CRUD*
Sea Acceptance Trials		*CRUD*	*R*
Ship Performance Assessment - ME		*CRUD*	*CRUD*
Ship Performance Assessment - WE		*CRUD*	*CRUD*
Ships Log		*CRU*	*CRU*
Spectrometric Oil Analysis Program (SOAP)	*R*	*CRUD*	*CRUD*
Standard Operating Checks		*CRUD*	*CRUD*
Survey Reports - Dockyard. + OEM	*CRUD*		*CRUD*
Test Specification & Results	*CRUD*	*R*	*R*
Thermography	*CRUD*	*R*	*R*
Unit Maintenance Management Systems	*CRUD*	*CRUD*	*CRUD*
Variation Orders	*CRUD*		
Vibration Monitoring & Analysis	*R*	*CRUD*	*CRUD*
WE Defect Log		*CRUD*	*CRUD*
Work Requisition Form	*CRUD*		

All the information sources either provide a degree of detail regarding the material state of a system or information that supports or disprove a hypothesis. For example, the ships log [22] not only details the location, course, etc. of a vessel but also speed, sea state, air / sea temperature, distance run; environmental factors that may influence system degradation / defects. Information contained within the ships log (Table 3) is a combination of objective and subjective data providing context to succinct preventive / corrective maintenance records.

A review of the data operations within each cyclical phase indicates the lifecycle of the information source, utilisation and potential value of the information.

Table 3: Objective / Subjective

Type	Ships Log
Objective	At 18:00 air temp. 39°C and sea temp 32°C [23]: high ambient temperatures will reduce the efficiency of chilled water systems essential for cooling weapons and sensors
Subjective	"Ice lookouts were relaxed. [...] The ship was in very heavy weather, some securing of the foc'sle required. Maintenance all over the ship" [24], throughout the day the vessel was also in moderate (4~8ft) to rough (8~13ft) sea at an average speed of 11½ knots.

Three information sources in Table 2 are seen to be maintained throughout each cyclical phase, i.e. Material Usage, OpDef's and UMMS. Combining the information could potentially provide an indication of the material state, spares / material usage, preventive and corrective maintenance undertaken and defects that degrade the operational capability of the vessel. It should be noted, where combining data is not seen to meet the functional requirement of indicating the material state, new / additional data sources may need to be developed.

The manner in which systems are utilised can skew any potential knowledge that might be inferred, e.g. prior to deployment a vessel may endeavour to draw spares for the period away. Subsequent analysis would possibly indicate high usage and hence infer a potential problem; the data must therefore be seen in the context of a vessel's overall programme. Furthermore, the management of onboard systems such as UMMS can vary markedly with respect to the frequency, volume and type of information recorded.

As stated above, information may be seen to exist in two domains, i.e. onboard and onshore, however, as a consequence of operational and / or software functionality constraints, Common Information may encounter latency or cease to exist, e.g. a Vanguard class submarine will undertake a 3 month submerged patrol with no contact with onshore support (Figure 6).

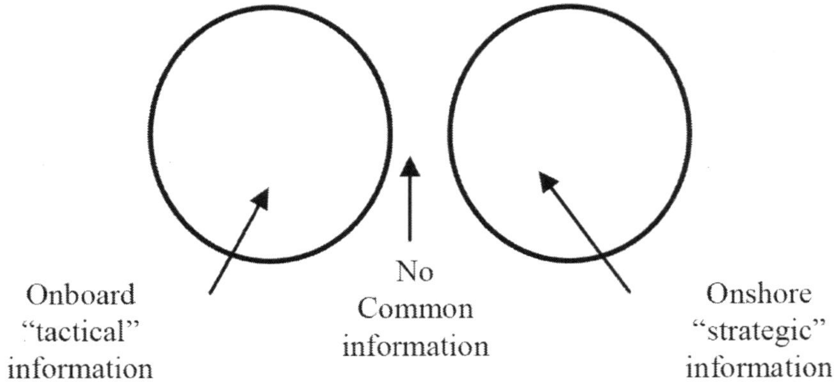

Figure 6: Operational Information Limitations

Normally, the minimum Common Information interface will comprise of OpDef signals. In the absence of a communications link, the vessel will function as an autonomous entity and re-establish the Common Information to onshore authorities when possible.

Upkeep information sources may be divided into three functional domains (Figure 7), i.e.

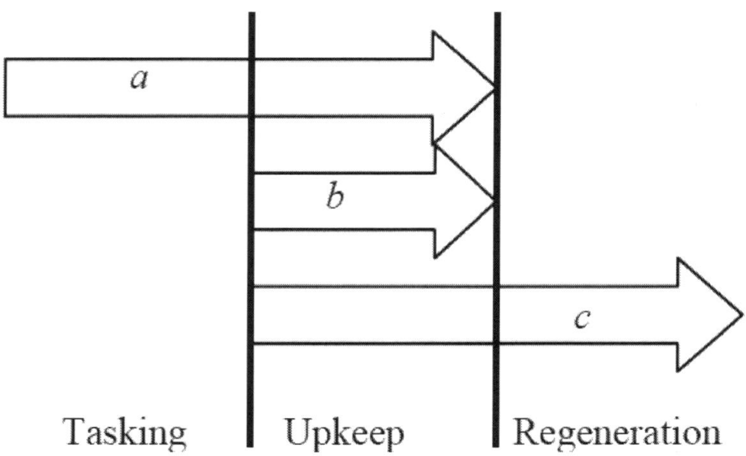

Figure 7: Upkeep Information

"*a*" Created during Tasking but utilised within Upkeep Surveys provide a high level of detail with respect to the condition of systems / vessel, enabling early identification of areas of degradation / defect, facilitating early ordering of long lead materiel. Other sources include condition based data, providing a perspective of the condition of systems whilst operational, e.g. vibration, SOAP, fresh water coolant.

"*b*" Created and utilised within Upkeep

Information sources created only for the Upkeep phase, e.g. Variation Orders, A's & A's, End of Project Report, Work Requisitions. May be cconsidered "project" based, i.e. focused upon delivering and completing the project. The stakeholder is typically the maintenance organisation.

"*c*" Created during Upkeep and utilised later

Information created during Upkeep, providing a datum for future reference, e.g. Test results, MSP, typically these enable stakeholders to assess degradation or validate / verify the work undertaken.

The three functional domains reflect the changing involvement and interest of the stakeholders, e.g.

"*a*" Created during Tasking but utilised within Upkeep Customer: COM; DE&S

Created by: RN, Maritime Capability Trials and Assessment; maintenance organisation; OEM

"*b*" Created and utilised within Upkeep

Customer: COM; DE&S, maintenance organisation Created by: maintenance organisation; OEM

"*c*" Created during Upkeep and utilised later Customer: RN; DE&S

Created by: COM; maintenance organisation; OEM

In the absence of a cost benefit analysis the "seven laws of information" provide a structure when assessing the value of information. Following the interviews and appraising the data, each information source has been reviewed against each "Law", whereby, High (*H*) = highly relevant / applicable, Medium (*M*) = moderately relevant / pertinent, Low (*L*) = low applicability / relevance. Table 4 details each information source and an assessment to quantify the value of each information source.

The applicability of a number of the Laws are fixed, e.g. "Information is not Depletable" – unlike a natural resource which may be depleted with use, information remains extant and may be used repeatedly without depleting the source. Consequently, all information sources are shown as "*H*" ~ High. Similarly, "The Value of Information Increases with Accuracy" may also be considered true; the antithesis would be it "increases with inaccuracy". Inaccurate data may be resolved by mathematical techniques [25], however, the preference would be accuracy, thus enabling decision making.

The general premise is "more is better", e.g. natural resources, however, with respect to data the construct "More Is Not Necessarily Better" may be valid. Large volumes of data may necessitate different techniques [26], e.g. data variances from the norm, allowing the system to identify change. Consequently, the default for each information source is "*L*" ~ Low. A number of information sources are null, e.g. FLOC, these sources may contain confidential information which would be inappropriate to assess in this paper

A number of information sources are consistently "*H*" ~

High, typically they are,

- Condition Based Data: e.g. Fresh Water Coolant, Fuel Analysis, Vibration monitoring, each provide information considered "high" value due to the immediacy, objective content and relevance to the condition of the system.
- Engineering Logs: the logs provide a high level of immediacy and relevance to the condition of the system. The data is highly perishable given the requirement to sustain and maintain systems.

- Surveys: considerable detail and accuracy with respect to the condition of the hull and systems.

A common characteristic of the information sources identified is the objective nature of the data contained within, e.g. Condition based data ~ vibration analysis (Figure 8).

Table 4: Application of the "Seven Laws of Information"

	Information Laws						
Information Source	1	2	3	4	5	6	7
Alteration's and Addition's (A's & A's)	*M*	*M*	*L*	*H*	*H*	*L*	*H*
Certificates, e.g. Naval Ship Safety Cert.	*H*	*H*	*H*	*H*	*M*	*M*	*H*
Dynamic Machinery Trial	*H*	*H*	*H*	*H*	*H*	*L*	*H*
End of Project report		*H*	*M*	*H*	*H*	*L*	*H*
Final Signal (FINSIG)			*M*	*H*	*H*	*L*	*H*
Financial reports			*L*	*H*	*M*	*M*	*H*
Fleet Operational Capability							
Force Generation Order							
Fresh Water Coolant	*H*	*H*	*H*	*H*	*H*	*L*	*H*
Fuel Analysis	*H*	*H*	*H*	*H*	*H*	*L*	*H*
Fuel and Lubrication Report (FLUBCON)	*L*	*H*	*M*	*H*	*H*	*L*	*H*
Harbour Acceptance Trials		*M*	*M*	*H*	*M*	*L*	*H*
Lloyds Survey	*H*	*M*	*M*	*H*	*H*	*L*	*H*
Logistics Log	*H*	*H*	*M*	*H*	*M*	*M*	*H*
Material State Portfolio		*M*	*L*	*H*	*H*	*L*	*H*
Material Usage	*M*	*M*	*M*	*H*	*H*	*L*	*H*

ME Defect Log	*H*	*H*	*H*	*H*	*H*	*L*	*H*
Night Rounds Report	*H*	*H*	*H*	*H*	*H*	*L*	*H*
Non-Destructive Testing, e.g. X-Ray, Ultrasound	*M*	*M*	*M*	*H*	*H*	*L*	*H*
Operational Capability Confidence Check							
Operational Defects							
Pre-Upkeep Material Assessment	*M*	*H*	*H*	*H*	*H*	*L*	*H*
Pre-Upkeep Weapon System Assessment	*M*	*H*	*H*	*H*	*H*	*L*	*H*
Ready For Ops Signal							
Report of Machinery Trials in Ships	*M*	*H*	*M*	*H*	*H*	*L*	*H*
Risk Register	*H*	*H*	*M*	*H*	*H*	*L*	*H*
S2022's Defects	*M*	*M*	*M*	*H*	*H*	*L*	*H*
Sea Acceptance Trials							
Ship Performance Assessment - ME							
Ship Performance Assessment - WE							
Ships Log	*M*	*M*	*L*	*H*	*M*	*L*	*H*
SOAP	*H*	*H*	*M*	*H*	*H*	*L*	*H*
Standard Operating Checks	*M*	*H*	*H*	*H*	*H*	*L*	*H*
Survey Reports - Dockyard. + OEM	*H*	*H*	*H*	*H*	*H*	*L*	*H*
Test Specification & Results	*H*	*M*	*M*	*H*	*M*	*L*	*H*
Thermography	*H*	*H*	*M*	*H*	*H*	*M*	*H*
UMMS	*H*	*H*	*M*	*H*	*H*	*M*	*H*
Variation Orders	*H*	*H*	*M*	*H*	*H*	*M*	*H*
Vibration Monitoring	*H*	*H*	*M*	*H*	*H*	*L*	*H*
WE Defect Log	*H*	*H*	*H*	*H*	*H*	*L*	*H*
Work Requisition Form	*H*	*H*	*M*	*H*	*H*	*L*	*H*

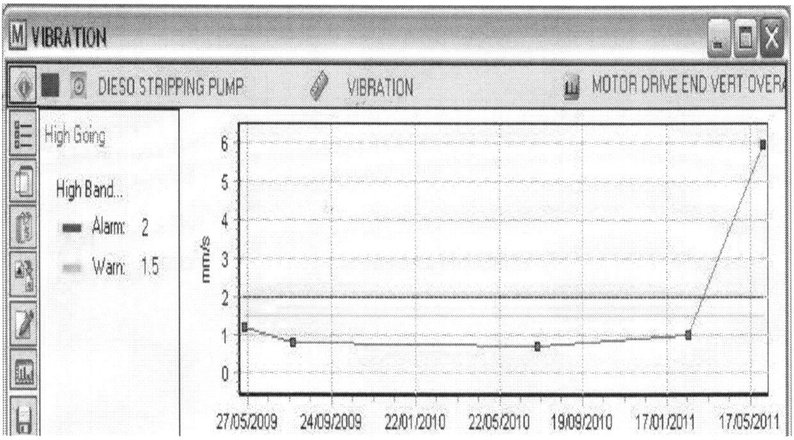

Figure 8: Vibration Analysis

There are a number of differences, e.g. condition based data is typically structured, e.g. running hours, whereas engineering logs are often unstructured handwritten records (Figure 9). A key advantage of processing objective data is the application of comparatively simple logic, whereas subjective data frequently requires interpretation, time, experience, training, etc.

Each information source will exhibit properties, e.g. timeliness, accessibility, standards, similarly each stakeholder will define a requirement, e.g. Marine Engineering (ME) Defect Log, the RN ship staff may require low latency, abstraction and be structured. However, the information source may exhibit minimal latency, abstraction, but is highly unstructured. Figure 10 illustrates the variances in properties and user requirement.

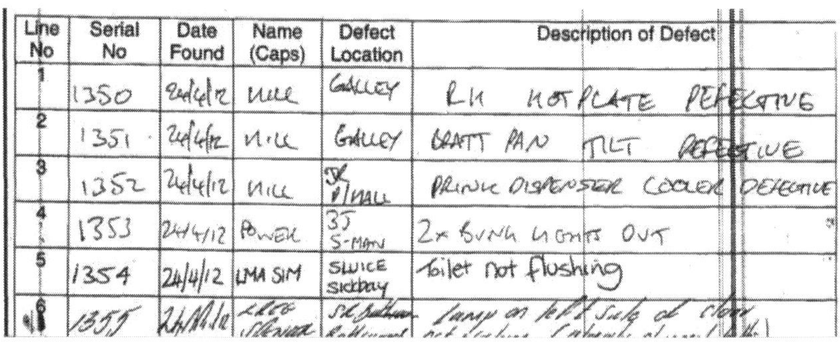

Line No	Serial No	Date Found	Name (Caps)	Defect Location	Description of Defect
1	1350	24/4/12	NILL	GALLEY	RH HOT PLATE DEFECTIVE
2	1351	24/4/12	NILL	GALLEY	BRATT PAN TILT DEFECTIVE
3	1352	24/4/12	NILL	3K MALL	DRINK DISPENSER COOLER DEFECTIVE
4	1353	24/4/12	POWELL	3J S-MAN	2x BUNK LIGHTS OUT
5	1354	24/4/12	LMA SIM	SLUICE Sickbay	Toilet not flushing
6	1355	24/4/12	[illegible]	[illegible]	[illegible]

Figure 9: Unstructured Marine Engineering Defect Log

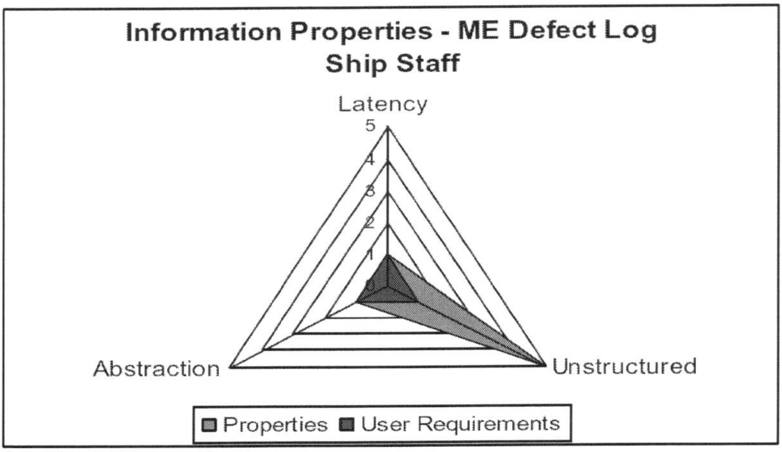

Figure 10: Unstructured ME Defect Log – Stakeholder – Ship Staff

The same information source may exhibit different properties for different stakeholders, each with their own user requirements, e.g. Figure 11 illustrates the same information source for the maintenance dockyard.

Hence, for a given information source there is a marked variance in the properties and user requirement; reflecting differences in the information domains, i.e. onboard / onshore.

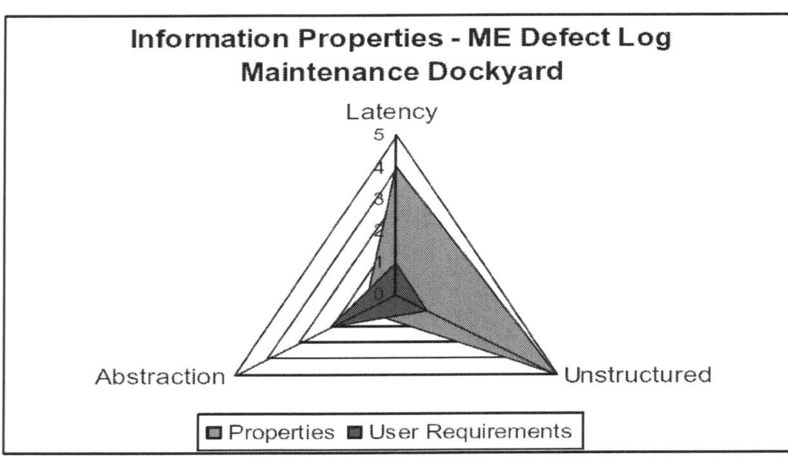

Figure 11: Unstructured ME Defect Log – Stakeholder – Maintenance Dockyard

5. DATA ABSTRACTION AND UTILISATION

The information sources discussed may be seen to exist at varying levels of abstraction (Table 5), i.e.

Table 5: Levels of Information Abstraction

Level	Information	Abstraction
a	Operational / Command: high level of abstraction, providing summary of system availability, capability or indication of severe defect / failure, e.g. OpDef's, FLOC	Most
b	Functional: moderate level of abstraction, managing the maintenance task, e.g. UMMS	↕
c	Physical: condition based data, real time, e.g. SOAP, vibration monitoring	Least

Information may exist at all three levels, however, depending upon the cyclical phase, the level of utilisation will vary (Figure 13, 14), e.g.

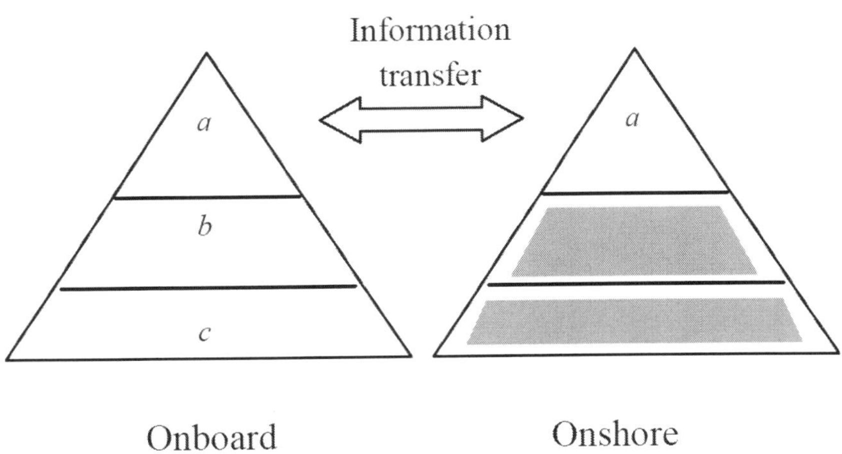

Figure 13: Tasking / Regeneration Information Utilisation

During the Tasking / Regeneration phase a high level of abstraction may be seen to exist with respect to "information transfer" for onshore authorities. Onshore authorities will typically require a higher level of tacit knowledge to formulate the lower levels of information.

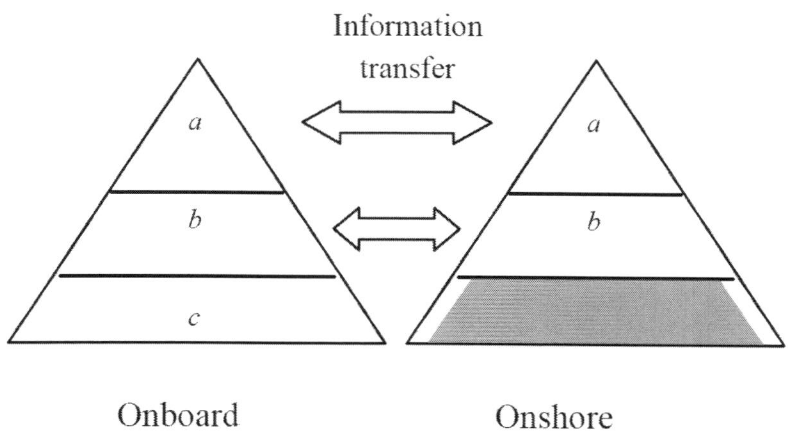

Figure 14: Upkeep Information Utilisation

During Upkeep, lower levels of information may be available and utilised; however, the availability and content of the lowest information level, i.e. "*c*" may vary. Where data is not available tacit knowledge to retrospectively deduce the missing information is often required. The lack of information at the lowest level may be a consequence of numerous systemic factors, including, latency, lack of retention of onboard data, format of data: e.g. paper not digital.

6. CONCLUSION

The application of the "seven laws of information" provides a structure to assess the value of each source. However, the validity and applicability of each "law" will vary with each source, e.g. it would not be appropriate to apply Law 1 "Information Is (Infinitely) Shareable" to highly classified data. The technique detailed by Moody

& Walsh [18] focus's upon the service / financial potential of the information source. Hence, a source identified consistently as high may be perceived as high value, enabling maintainers "to achieve higher order organisational objectives and generate capabilities that create sustainable advantage" [15].

A number of information sources exist and are utilised throughout each cyclical phase, i.e. UMMS, OpDef's and material usage offering the potential to bestow the greatest value with respect to indicating the material state. However, variations in local management of onboard systems may negate the potential value. The information contained within condition based data, engineering logs and surveys are also identified as possible high value sources. Within the Tasking phase this information may also be considered valuable. However, within the Upkeep phase the value may be diminished as a result of latency and abstraction. The consequence of abstraction includes aggregation and loss of data / detail, factors which potentially degrade the provision of timely and effective maintenance. Abstraction may be minimised by maintaining and retaining low level records, e.g. electronic data logging of condition based data, utilising computer systems, e.g. UMMS, rather than create paper based systems, data logging of environmental data, the collected data would then be available for diagnostics and potentially prognostics.

The focus of information assessment is primarily retrospective, i.e. Upkeep will review the performance and condition of systems in Tasking. However, a review of issues encountered in subsequent phases, e.g. Regeneration may provide an indication of potential shortfalls during Upkeep, e.g. as detailed in the FINSIG.

Systems do not exist in isolation, hence a broader perspective is required, i.e. *Weltanschauung*, hence utilising high value data, e.g. condition based data, Ship Performance Assessment, etc. may provide that perspective.

Given the extended life of many vessels and the considerable time undertaking Tasking utilising data from disparate sources, e.g. ships log, Marine / Weapon engineering logs, Surveys, etc. will help to create a "richer picture". As defined by Law 5, "The Value of Information increases when combined with other information", thus a more accurate material state may be established by combining information to improve maintenance decision making and reduce emergent and unplanned work.

ACKNOWLEDGEMENTS

The work reported was supported by the Bristol/Bath Industrial Doctorate Centre in Systems, funded by EPSRC grant EP/G037353/1.

REFERENCES

1. INCOSE, 2010. Systems Engineering Handbook. A Guide for System Life Cycle Processes and Activities. Ver. 3.2 INCOSE, San Diego, CA. January/2010.

2. Murthy, D.N.P., Asgharizadeh, E., 1999. Optimal decision making in a maintenance service operation. European Journal of Operational Research 116 (1999) 259-273

3. Kelly, E., Ratchev, S., 2011. Integrated Approach to Maintenance and Capability Enhancement: Proposal of a Maintenance Dashboard Framework. Springer London. DOI 10.1007/978-0-85729-189-9_19

4. Hobson, B., 2008 Obsolescence Challenges and the Canadian Navy. Canadian Naval Review. Summer 2008.

5. Roulston-Eldridge, J, 2014. HMS Illustrious Port Outer Gas Generator Exchange The Last Mount of Olympus at Sea? The Naval Engineer. Spring 2014

6. MOD-AOF2, 2013. Acquisition Operating Framework. [ONLINE] Available at: https://www.aof.mod.uk/index.htm [Accessed: 12/Feb/2014]

7. MOD-AOF3, 2013. The CADMID Lifecycle. [ONLINE] Available at: https://www.aof.mod.uk/aofcontent/strategic/guide/lifecycles/sg_cadmid .htm?zoom_highlight=cadmid [Accessed: 10/Nov/2013]

8. Ward, R.C., Loftis, J.C., McBride, G.B.,1986. The "data-rich but information-poor" syndrome in water quality monitoring. Environment management. Volume 10, Issue 3, pp 291-297

9. Slotnik, W. J., Orland, M., 2010. Data rich but information poor.*Education Week.*

10. Timmerman, J, G., Beinat, E., Termeer, K., Cofino, W., 2010. Analyzing the data-rich-but-information-poor syndrome in Dutch water management in historical perspective. Environmental management, 2010, Vol.45(5), pp.1231-42

11. Forte, P., 1994. Data rich, information poor: Data, information and decision support in the NHS. *European Journal of Information Systems*, *3*(2), 148-154.

12. Bouma, J.A., van der Woerd, H.J., Kuik, O.J., 2009. Assessing the value of information for water quality management in the North Sea. Journal of Environmental Management 90 (2009) 1280–1288

13. Fritz, S., Scholes, R.J., Obersteiner, M., Bouma, J., Reyers. B., 2008. A Conceptual Framework for Assessing the Benefits of a Global Earth Observation System of Systems. IEEE Systems Journal, Vol. 2, No. 3, September 2008

14. Braganza, A., 2004 Rethinking the data–information–knowledge hierarchy:towards a case-based model. International Journal of Information Management 24 (2004) 347–356 HMG Security Policy Framework. Version 11.0 – October 2013

15. Joint Doctrine Publication 0-10 British Maritime Doctrine. August 2011. Swindon, Wiltshire: Ministry of Defence

16. Moody, D., Walsh, P., 1999. Measuring The Value Of Information: An Asset Valuation Approach. European Conference on Information Systems (ECIS'99)

17. Weber, C., Deubel, T., 2003. New Theory-Based Concepts for PDM and PLM. International Conference on Engineering Design. August, 2003

18. Gavin, A., Kelly, C., Nic Gabhainn, S. and O'Callaghan, E. (2011) *Key issues for consideration in the development of a data strategy: A review of the literature.* Dublin: Department of Children and Youth Affairs. Available at: www.dcya.ie

19. Material State Portfolio for HMS . 2013

20. Final Signal (FINSIG) Extract for HMS. 2013

21. RN Form S322 Ship's Log. Revised 2/01. Ministry of Defence

22. Ships Log HMS. 2009

23. Ships Log HMS. 2010

24. Blanco, L., Crescenzi, V., Merialdo, P., & Papotti, P. (2010, January). Probabilistic models to reconcile complex data from inaccurate data sources. In *Advanced Information Systems Engineering* (pp. 83-97). Springer Berlin Heidelberg.

25. Tien, J.M., 2013. Big Data: Unleashing information. Journal of Systems Science and Systems Engineering. June 2013, Volume 22, Issue 2, pp 127-151

CHAPTER 6

Naval architecture, expertise and navigating authority in the British Admiralty, *c.*1885–1906

Don Leggett*

School of History, University of Kent, Canterbury, UK

ABSTRACT

Naval architecture and the ship design process provide useful windows into the cultures of authority and expertise at the British Admiralty. Naval architecture was a complex field of theory and practice that underwent a number of changes in the nineteenth century, from the technical to the professional. It was also a highly contested and controversial subject, in which practitioners from the naval service to civilian engineering sought authority. This article provides a survey of naval architecture, authority and expertise, before proceeding to two case studies that reveal important changes to the boundaries of technical expertise in the ship design process in the late Victorian and Edwardian Admiralty.

KEYWORDS

authority, expertise, HMS *Dreadnought*, HMS *Royal Sovereign*, John Fisher, naval architecture, Royal Navy, William H. White

In 1889, William H. White, Director of Naval Construction (DNC) at the Admiralty, presented the design for the Royal Navy's new class of *Royal Sovereign* battleships in a paper at the Institution of Naval Architects (INA). This was an unusual occasion, as it was rare for the INA to host papers containing sensitive material from a national security and secrecy perspective. White had asked his political superiors for this opportunity in order publicly to respond to criticisms of the *Royal Sovereign*'s design. The main antagonist was the retired Chief Constructor of the Navy, Edward Reed. Now the MP for Cardiff, Reed wrote via *The Times* that there had been a number of 'rash and dangerous innovations' in ship design. He was especially concerned about the sacrifice of armour coverage for greater seakeeping and offensive qualities.[1] At the INA White received tough questions from Reed, but also generous support from the naval officers in the audience. Rear-Admiral Philip Colomb expressed his support for White and the Admiralty's decision to change the design emphasis from defence to offence: 'we build these ships ... to gain the victory'. Vice-Admiral Geoffrey Hornby, President of the Royal Naval College, praised the *Royal Sovereign*'s promised speed, telling members that it was 'the highest quality that a ship can have'. He supported this by citing 'the opinion of our highest authority, that of Lord Nelson', who he believed favoured great speed in his choice of ships. The accuracy of the claim aside, Nelson's authority on naval warfare continued to resonate almost a century after Trafalgar, and his name was a powerful resource to muster.

Another naval officer, Captain Charles Beresford, added his approval for the design, explaining that naval officers needed the tools to attack their opponents rather than defend against them.[2] He was also pleased with how the Admiralty had gone about the design of the *Royal Sovereign*. In August 1888, the First Lord of the Admiralty, George Hamilton, brought together a number of senior naval officers and the DNC at Devonport Dockyard to discuss future warship construction. The First Lord convened this meeting 'for the guidance of the Director of Naval

Construction in preparing alternative outline designs'.[3] The Admiralty Board provided a set of design requirements for White to fulfil, mostly concerning armaments. Rather than requesting that a detailed list of specifications be met, the board wished to see a variety of proposals that met the specified requirements. On 16 November the same individuals met again in the First Lord's rooms at the Admiralty to discuss White's designs. Back at the INA Beresford praised the 'business-like way' these ships had been conceived: 'They have got the seamen together. ... They put down what they wanted, and then sent for a naval architect, and said: With that buoyancy, with that draught to carry such and such a weight, which includes the whole armament and weighting, what will you design?'[4]

Beresford's praise for the design process employed for the *Royal Sovereign* was tempered by his perception of how the Admiralty had historically undertaken design work. He explained:

What I have objected to for many years has been what are called naval constructors' ships. My opinion is this: I say that the way you have built your ships up to date has not been fair on the naval constructor, any more than it has been on the seaman. It has been a haphazard way of doing it. ... Many of these ships, which Sir Nathaniel Barnaby and others have been abused for building, were not built with any definite idea of the weight which they were to carry.[5]

Years of controversy, public accusations and antagonism over ship design had polarised debates between naval officers and architects in the second half of the nineteenth century.[6] From Beresford's perspective, naval architects had been given a broad grant of authority to provide the design aims for the navy's battleships, largely due to a lack of naval supervision. By his reckoning, constructors 'never consulted the seamen ... [but] the man first to consult is the man who has to go to sea in her and to fight in her, and then the man who is to build her should ascertain how far he can meet the requirements of the man who has to fight her'.[7]

Beresford's perception of how the design process historically worked was not particularly accurate, as the next section of this article demonstrates. That should not rob his comments of significance. The public claims that naval officers and architects made were part of a process of contesting authority and seeking recognition. For example, Edward Reed painted a very different, conflicting picture of the design process. He contended that naval architects lacked the authority to shape designs, even lacking control in such a fundamental way as guaranteeing ship safety. At the same 1889 meeting of the INA he alleged that the sacrifice of armour for fighting qualities in the design of the *Royal Sovereign* was the product of naval pressure on White: 'I have not the slightest doubt that he is obliged to stand idly by and take all the guns and other conditions which he says the Board of Admiralty impose upon him.'[8]

This was a continuation of a line of argument about the authority of naval architects that Reed had been making since the 1860s. He believed that naval architects were constrained by officers, whose ignorance of the scientific principles at work in ship design endangered the navy. He believed that there was a fundamental difference in how the two groups saw ship design, and sought to condemn naval officers for flagrantly dismissing the scientific authority that naval architects possessed. As he wrote in the pages of the *Annual of the Royal School of Naval Architecture and Marine Engineering* in 1872:

The whole strength of the scientific advisers of the Admiralty, from the date of the *Warrior*'s design to that of the *Captain*'s, was exerted in favour of high-sided, seaworthy ships, and against Monitors and dangerous vessels of every form. At length the action and the agitation of the unscientific begat so much confidence, even in the minds of the highest political and naval authorities, that some concessions became imperative, and there is reason to fear that but for the (in one sense timely, although in another sense untimely) loss of the *Captain*, men of science would have been completely overruled, and the future ironclad Navy would have been constructed in violation of scientific principles.[9]

Thus Reed described the controversy between naval officers and naval architects as a contest where political and naval authority was above scientific authority. His 1889 remarks at the INA continued this argument, making the case that the expertise of naval architects was frequently trampled upon by politicians, administrators and naval officers.

These discussions at the INA reveal some of the ways in which ship design was an important battleground in the history of authority and expertise within the Admiralty. With this established, this article has two primary objectives. The first is to provide a brief history of this battleground, highlighting some of the ways that the issues pertaining to authority and expertise crop up in the institutional, professional and political history of naval architecture. The second is to reconstruct how authority was navigated in the ship design process specifically, charting the changing boundaries of the naval architect's expertise. This is done across two case studies, focused on the design meetings that preceded the construction of the *Royal Sovereign* (1891) and HMS *Dreadnought* (1906) respectively.

The metaphor of navigating authority provides an antidote to a recurring theme in naval history. All too often, studies treat expertise and authority as interchangeable concepts, rarely asking how experts (particularly in emergent areas of expertise) gained authority. This question is incredibly important for understanding how naval architects associated with the INA gained influence and power within the Admiralty to practise the engineering science that was so central to the design of powerful battleships in the closing decades of the nineteenth century. In *Warrior to Dreadnought* (1997) – often treated as authoritative by naval historians – David K. Brown treats changes in the practice of naval architecture as a simple shift from 'rules of thumb' to 'calculation and model testing'. He places credit with Reed, 'who himself made important contributions but also encouraged his staff and both accepted and sponsored work from outside the Admiralty', but makes nearly no mention of how Reed gained authority. What little is noted about Reed's dealings with politicians highlights how fraught his relationships were, and his lack of credibility.[10] John Beeler's *Birth of the battleship* (2001) similarly fails to trace the limits of authority that naval architects possessed, focusing instead on the politics of ship design at the level of the Board of Admiralty, parliament and public discourse.[11] This is not unusual in naval history, and indeed there are a number of studies that treat naval architects as pawns in the politics of naval power that rarely move from examining partisan politics to the politics of expertise and authority.[12] The rise of the scientifically trained naval architect within the Admiralty was neither inevitable nor straightforward. They competed for authority with both naval officers, whose judgement of ship design and behaviour was honed at sea, and shipwrights, who trained as apprentices in the craftwork systems of the Royal Dockyards.

1. AUTHORITY ACROSS POLITICS, PROFESSIONS AND INSTITUTIONS

Naval ships were products of many types of expertise. From a scientific and engineering perspective, they involved a huge amount of specialised knowledge and skill, including mathematics, mechanics and various woodworking crafts (and later metallurgy). From a naval perspective, they represented the outcome of strategic and operational thinking. Their designs embodied the expertise that naval officers obtained over a career of studying naval battles and serving the Royal Navy at sea. These types of expertise were not always complementary, as naval architects and officers came into conflict. Moreover, these tensions were frequently politicised by the involvement of politicians, partisan politics and parliamentary scrutiny. The following account of naval architecture highlights these and other concerns relating to the function of authority and expertise in this field.

The politicisation of naval architecture was perhaps clearest to see in the appointment of Admiralty officials. In 1832 William Symonds, a naval officer, was appointed Surveyor of the Navy, above the navy's senior shipwrights and the graduates of the School of Naval Architecture. He neither possessed training in naval architecture nor experience working in the Surveyor's office or the naval dockyards. He did, however, have limited experience designing a small number of yachts and smaller naval vessels.[13] Why, then, was Symonds put in charge of naval ship design and construction? The Whig Admiralty that held sway during the passage of the Great Reform Act saw an excellent opportunity to remove Tory influence from the navy and naval constituencies. But the politics of reform and the appointment of Admiralty officials were connected in more complex ways. As Andrew Lambert has examined, Symonds's powerful patrons, the Duke of Portland and Lord Vernon, were important allies for the Whigs as they sought to pass the Great Reform Act.[14] This episode highlights an important issue in the history of expertise common to government work: politicians with little specialised knowledge frequently judged between the expertise of naval officers and architects. In Symonds's case, the appointment was made by the Prime Minister, Earl Grey, and the First Lord of the Admiralty, James Graham. Unable to judge the merits of potential appointees in any sophisticated manner as equals (by possessing expertise in naval architecture), they examined credentials, employed patronage and made decisions based on ulterior motives.

Partisan politics were not the only type of politics at play in Symonds's appointment. This episode also highlights the importance of social and institutional authority. Earl Grey urged Graham to take firm control over the navy's dockyards and shipbuilding work.[15] The First Lord hoped that appointing Symonds would help him achieve that: 'I consider that the application and economy of labour, the discipline of the dock-yard, and the adjustment of the whole scheme of labour, are all duties intimately connected with the office of surveyor.'[16] Graham did not believe that this would be possible were a shipwright appointed Surveyor: 'A person who has himself been a shipwright, would necessarily have too close a connexion with the other shipwrights of the yard, to enable him to exercise that control of them which the public interest may make it requisite should be exercised.'[17] Graham did not believe shipwrights were capable of taking control over the dockyard because they belonged to the same artisan class as those they would have to manage.

The navy's shipwrights were unhappy with Symonds's appointment. In 1833 Henry Chatfield, a shipwright and graduate of the School of Naval Architecture, published a pamphlet attacking the decision:

When it is thus stated before the Representatives of the Nation, that a whole body of public servants, professionally educated, are less competent to fill an office of responsibility in their own department than an amateur ship-builder, it becomes a serious charge indeed; and they whose reputation is thus involved, are loudly called upon, by every feeling which ought to actuate honourable minds, to endeavour, by proper means, to exonerate themselves from so humiliating a reflection.[18]

Chatfield claimed that the authority that master shipwrights possessed was rooted in their education and professional service, while Symonds's authority was an extension of political power. This was a prescient insight, as Symonds only remained in office for so long as his political patrons Portland and Vernon maintained influence over the Admiralty. When that ceased to be the case he resigned, and was replaced in 1847 by Baldwin Walker.

Walker's appointment marked a change in the responsibilities of the Surveyor's office. The First Lord, the Earl of Auckland, explained to Walker: 'I would require in future more of judgement and general superintendence and much less of minute interference in construction than have yet been expected in this office.'[19] Symonds had not formally been required to design ships, but in practice had taken a leading role in this work. The board made clear to Walker that although he was to be the final arbiter in his office, he was to use his two assistants to draw up ships' plans. Under Walker, the management of naval architecture became far less personally controversial than it had under his predecessor. His period in office also demonstrates how institutional reforms could curtail the authority of professional offices within the Admiralty. Auckland, in judging that naval officers ought not to design ships but did possess the right skills and experiences for the management of shipbuilding, altered the boundaries of what the Surveyor could do.

Walker was largely welcomed by the navy's shipwrights, but he and his brother officers fast became a target for men of science and naval architects who wanted to extend the influence of a scientific approach to ship design. In the early 1860s William Snow Harris produced a pointed critique of the navy's management approach and treatment of technical specialists. A fellow of the Royal Society and a trusted adviser to the Admiralty for his work on lightning conductors, Harris was not easy to dismiss. 'It is certainly a very great mistake', he wrote, 'to place the Naval Construction of this country under the control of a Naval Officer or any other individual, not fitted by a sound mathematical, physical, and practical education for so great an object.'[20] A second, more influential attack came from the naval architect John Scott Russell. Russell had no formal affiliation to the Admiralty, but he had collaborated with Isambard Kingdom Brunel on the SS *Great Eastern* (contributing the hull form, paddle wheel engine design and overseeing construction) and the Admiralty on ironclad shipbuilding, notably the *Warrior*.

Russell had a professional stake in the authority of naval architects, and in 1861 produced a pamphlet on the politics of ship design. It contained sections on 'Iron against wood – the American battle', 'Who is to construct the navy? – The Admiralty – or the House of Commons?' and 'How to find the money; or, Fleets *v.* Establishments'. Russell made a strong argument for taking control over design work out of the hands of politicians and administrators. He was specifically concerned by the accountability of government representatives, arguing that they were more receptive to discussions in parliament than the concerns of specialists who possessed knowledge and experience in ship design. In short, that they had a greater obligation to public opinion than expert opinion.[21] Russell believed that this was a serious obstacle to the rule of technical specialists in the Admiralty. Again, the political theme emerged as an obstacle to authority. The best solution Russell saw was for the Admiralty to cease its own design work, and instead purchase its ships from commercial builders such as himself.[22]

Russell did not simply diagnose problems relating to authority, he offered the means to remedy them. In 1860 he brought together Reed, Nathaniel Barnaby and Joseph Woolley to form the Institution of Naval Architects. The INA became a forum where scientific ideas could be shared, results of experiments discussed and the lessons of experience interrogated. More importantly, however, the founders of the INA sought a new identity for naval architects. Reed in particular believed that the INA's greatest challenge was to mark out the ‘few real naval architects in the country’, and promote their expertise within the naval community.[23] Members of the INA broke from many of the craft-orientated and rule-of-thumb traditions in which shipwrights and naval officers worked through the problems of ship design. In their place they encouraged precision measurements, greater use of theory and more respect for scientific authority. They effectively sought to make the ship an object of scientific enquiry, and the ship designer a man with scientific credentials.

Through the INA a small network of naval architects, engineers, men of science and mathematicians grew. With the aid of Henry Cole and the Department of Science and Art, it successfully lobbied the government to establish a Royal School of Naval Architecture in Kensington (later relocated to Greenwich), away from the craftwork culture of the Royal Dockyards and the naval control that permeated them. Students of the school included William White, Philip Watts and Eustace Tennyson D'Eyncourt, all of whom went on to occupy the office of DNC at the Admiralty (as well as leading positions in warship design at Armstrong's Elswick shipyard). Meanwhile, INA members formed strong links with men of science and engineers like W.J. Macquorn Rankine, William

Thomson and William Froude, bringing the reputation of these men to bear on their work with the Admiralty.[24]

Naval architects were not the only group with a stake in ship design. Naval officers raised their own concerns about ship design in forums like the Royal United Services Institution and the letter pages of Britain's daily newspapers. Tensions grew between members of the two groups over the introduction of specific design features like heavy gun turrets. A focal point of this debate was the work of Captain Cowper Coles, a proponent of the turret. Coles was unhappy with how the Admiralty had undertaken experiments with this new armament, even going so far as to criticise the turret ship designed by Reed, HMS *Monarch* (1868). He utilised contacts in the navy and in parliament to lobby for a turret ship that met with his liking. By bringing political pressure to bear on the Admiralty he secured permission to work with the Birkenhead shipbuilder John Laird on a ship that he felt would demonstrate the qualities of a turret system, HMS *Captain* (1869).

The *Captain* was completed away from the supervision and oversight of Admiralty naval architects, but disaster struck during its trials in 1870. Over five hundred sailors lost their lives. The ship had been built heavier than its design specified, and the combination of low freeboard (the height of the deck above water) and poor stability proved to be a fatal combination.[25] In the subsequent investigation Reed was heavily criticised for supposedly not reporting his concerns over the ship's stability. The former Chief Constructor had a combative character, and continued a polemical discourse with politicians and naval officers after the disaster.[26] John Scott Russell took a different view of the *Captain*, painting it as another example of ship design driven by parliament and political authority rather than specialised knowledge.[27] While in an uncharacteristically unpartisan assessment of the disaster, an anonymous naval officer blamed the loss on a lack of understanding and trust between the navy's architects and officers.[28] In the aftermath of the *Captain* disaster naval architects solidified their claims to be the experts on ship stability, but larger design questions remained a source of contention. Writing on the political, economic, strategic, tactical and technological nexus at work between 1870 and 1881, John Beeler has argued that Reed's successor, Nathaniel Barnaby, did little designing: ‘He was not involved in the design of [HMS] *Shannon*, *Temeraire*, *Ajax*,

Agamemnon, *Colossus*, *Edinburgh*, *Conqueror*, and the 'Admiral' class at all.'[29] Changes to the DNC's institutional authority were, however, secured by the work of William White, serving first as Barnaby's assistant and then as his successor.

William White's campaign to extend the authority of naval architects at the Admiralty serves as a fine example of how an individual could secure greater authority by nurturing credibility and promoting greater understanding. White joined the Royal School of Naval Architecture in 1866, where he edited the school's journal, before proceeding to serve as Reed's secretary. White followed this appointment with service as an instructor at the school and then as assistant to Barnaby until 1883, when he was appointed manager of warship construction at Armstrong's. He returned to the Admiralty in 1885 as DNC, an office he held until 1902. While Reed attacked the opinions and authority of naval officers in the wake of the *Captain* disaster, White reached out to naval officers and engaged with their concerns in a way that did not antagonise.

White sought to promote greater understanding of what naval architects had to consider, and the pressures they had to balance, when producing a ship design. He wrote educational articles for the *Westminster Review*, *Fortnightly Review* and *Gentleman's Magazine*. He also undertook an ambitious programme to educate younger naval officers through a course of lectures at the Royal Naval College in Greenwich.[30] He believed that if he could shape the attitudes of young naval officers toward the science of naval architecture, then perhaps he could nurture a better understanding of the skills that constructors possessed and forge a stronger working relationship between officers and architects. Influential officers like Astley Cooper Key encouraged him to turn the lectures into a book, *A manual of naval architecture* (1877). Years later, in a letter to a naval officer, White explained that the book 'was written, as far as was possible, in popular language for the benefit of naval officers, yachtsmen and others'.[31]

White also sought to transform the institutional authority of naval architects, addressing both social and institutional aspects. The crux of this reform came in the formation of a Royal Corps of Naval Construction. David K. Brown, in his history of the corps, provides a narrow narrative on the professional formation of naval architects as managers. But reforming the design and construction processes of the Royal Navy involved complex and contingent problems concerning expertise: what knowledge, skills and authority should those in charge of the administration of Britain's techno-military complex possess? This was not a new concern, nor was it an issue on which only White had an opinion. Admiral Robert Hall, superintendent of Pembroke Dockyard from 1866 to 1871, argued that 'Being unconnected with any class in the dockyard, he [the naval officer] is readily acknowledged by all as their superior'.[32] Reed, in contrast, saw no reason why dockyard managers should be naval officers. The skills of a naval officer were best employed on a ship, while those of the engineer and naval architect were better deployed on land. In this split between land and sea, Reed sought to establish a 'natural' division of labour.[33]

Engineering a new system for managing the dockyards was no mean task. The organisation of labour in the dockyards continued along craftwork culture lines, with traditional structures for advancement. The existing system provided entry-level jobs for apprentices in assisting dockyard draughtsmen and dockyard foremen. Even graduates of the Royal School of Naval Architecture received no preferential treatment, unless they were appointed directly to the Chief Constructor's office. Crosbie Smith and Ben Marsden have observed that the engineering sciences were young professions, containing men who were overtly concerned with their 'social mobility on both personal and professional levels'.[34] This was certainly the case in naval architecture, and White believed that without the promise of professional advancement and social mobility, men of higher social standing would avoid the profession.

White devised a new scheme for the organisation and duties of the Constructor's department. His scheme rested on altering the process of promotion to reward scientific achievement, increasing the number of naval architects at the Admiralty to seventy-two, raising their status and expanding their role in the ship design, building and evaluation processes.[35] The key was to position naval architects as dockyard supervisors. These assistant constructors, as White termed them, would work under the Chief Constructor of the Navy, Nathaniel Barnaby, to provide a flow of information between the Constructor's department and the dockyards, while lending their knowledge and skills to dockyard officers. To gain credibility with naval officers serving in the dockyard, these assistant constructors would be required to go to sea for two years after finishing their studies in order to experience the behaviour of a ship, observe ocean waves and socialise more closely with naval officers. This reform would create a sphere of influence for naval architects across the entire shipbuilding process.[36]

White effectively redefined naval architects as managers and technical consultants. His rationale served his larger objective of establishing the useful roles that naval architects could take beyond design work within the Admiralty. As well as being experts of ship design, they would become recognised as expert mediators between their colleagues and ranking naval officers. Most significantly, the scheme would couple the recognition of the naval architect's particular skills with a grant of greater institutional authority. There would also be a change in title to reflect this managerial turn, with the Chief Constructor becoming the Director of Naval Construction (DNC). The corps was established on 23 August 1883. Shortly afterwards, White left Admiralty service to join Armstrong's warship yard at Elswick. He returned in 1885 as DNC, and in his tenure of office worked to continue extending the authority of naval architects in the Royal Navy by consciously nurturing credibility in their work among politicians and educating naval officers about design problems.

2. WILLIAM WHITE AND THE *ROYAL SOVEREIGN*

In August 1888 the First Lord of the Admiralty, George Hamilton, invited White to a meeting at Devonport Dockyard. Together with the Director of Ordnance, Captain John ‘Jacky’ Fisher, White joined a number of senior naval officers in a discussion on warship design.[37] Admiral John Hopkins, Controller of the Navy, recorded that Hamilton's aim in bringing together these people was to obtain the opinion of naval officers ‘in order that modern conditions might be most satisfactorily fulfilled’ in a first-class battleship.[38] To this end, the officers agreed on three key design requirements: that four heavy guns be placed in two protective stations at a considerable distance from each other and with an arc of training of roughly 200°; that a secondary armament be located in a central battery; and that the secondary armament be dispersed across two decks, to minimise the destructive potential of high-explosive weapons. White was then given leave to produce a set of designs that embodied these requirements. On 16 November the same individuals met again in the First Lord's rooms at the Admiralty to discuss the merits of the designs and ultimately choose one to be built.

Hamilton's meetings constituted a substantially different approach to the design work in which Barnaby had engaged during the 1870s and 1880s. It consisted of three reasonably distinct phases: consultation with naval officers about what the ship was required to do; design work undertaken by a naval architect, left free to act in accordance with the design requirements rather than detailed specifications; and a decision by naval officers guided by their naval experience.[39] In effect, the First Lord had formed a demarcation between the naval architect as designer and naval officer as judge, limiting their authority to specific stages in the design process. Previous interactions between naval architects and officers had neither been so close nor prescribed for a particular purpose, taking place through formal committee work, intermediaries within the Admiralty structure, like the controller, and informal meetings in learned societies.

In the first stage naval officers were encouraged to reflect on the arrangements of recent battleships. White reported that a lot of the judgements made about those ships at the meeting were based on the 'experience gained' from the various officers who served on them at sea. In particular, experience gained on the *Admiral* and *Trafalgar* classes suggested that the disposition of armour in those ships was superior to all others.[40] Other discussions at the meeting concerned the use of barbette gun mountings, which the DNC favoured.[41] Instead of the uniform thick armour provided by the citadel system employed in the *Inflexible*, which was advocated by the naval officer Arthur Hood, the barbette was an independent pear-shaped structure that presented a smaller target. It was still thickly armoured from top to bottom, particularly underneath the structure to protect against nearby explosions, but it remained a considerably lighter alternative. The meeting also saw discussions about defensive power and the desirability of greater seakeeping qualities than previous battleships. In the second design stage White was required to include barbettes in a way that met the officer's first design requirements. He was not, however, held to detailed specification about the ship's dimensions, weight and distribution of armour. Instead, White was free to develop a set of designs that incorporated the design requirements in a way that he felt the naval officers would appreciate – bearing in mind that he had been privy to their discussions. White personally favoured the barbette system, telling the INA the following year that it 'greatly economises weight of armour'.[42] He coupled this with a larger hull than was common in the navy, which meant that he was able to raise the ship's freeboard.

Lightness of armour, White recalled ten years later, was preferred by many naval officers if it provided better seakeeping qualities. Low freeboard could be beneficial in smooth waters, but in rough waters it diminished speed, stability and, in turn, the use of offensive power.[43] White was well aware of this in 1888, and his design for the *Royal Sovereign* ensured that seakeeping qualities were prioritised. This met with the approval of the naval officers. The report from the 1888 meetings describes the importance that was assigned to 'speed and fighting power at sea', and that 'the high freeboard and high placed guns in [the] design for [a] barbette ship ... were preferable to the lower freeboard and less height of guns in design for the turret ship of same displacement'.[44]

The design process for the *Royal Sovereign* allowed White a freer hand than previous DNCs and Chief Constructors, but at the same time restricted the role that he could play in setting design requirements and imposing his own views on the armour question on the Board of Admiralty. Describing this design process at the INA, White supported the limits that Hamilton had set him:

I would submit that it is for the naval architect to prepare alternative plans, showing possible dispositions of the protective material, within the limits of weight available in a design. Having done this, it is not his province, but that of naval officers, to determine from their knowledge of gunnery, what are the nature and extent of the damages likely to be done in action, and to choose that disposition of armour which they consider has the balance of advantage.[45]

White cut a conservative figure compared to Reed, who adamantly held to his conviction that ships should be designed by naval architects with as little interference from naval officers as possible. The former Chief

Constructor had long contended, with a tone of naivety and self-interest, that while engineers sought the ideal design, naval officers and administrators compromised that aim. He wrote in 1877 that 'many of the differences in our ships … [arise] from fancy, from caprice, from divided counsels, from the competition of influences within the Admiralty'.[46]

One important difference between White and Reed was that the former believed it was up to naval architects to judge what was safe concerning stability, and up to officers to judge what was appropriate for armour. Reed in contrast believed the naval architect was responsible for both. To this end, he criticised the decision to sacrifice armour for seakeeping qualities: 'the first consideration I think should be that that ship should be kept afloat by thick armour. … That has been the ground of the whole of my public agitation.'[47] Believing, naively, that naval architects favoured protection and naval officers firepower, Reed assumed that the *Royal Sovereign* represented a defeat for the authority of naval architects at the Admiralty. White took the opportunity in his INA paper to dismiss Reed's claims by revealing the negotiation of authority that lay in the decision to fit the *Royal Sovereign* with barbettes and a narrow armoured belt, 8 ft 8 in. wide (5 ft of which was below the waterline).

Reed's criticisms further establish the delicate balancing act that White performed as DNC. Ship design entailed compromises that required technical and social negotiations. In 1906 White reflected on this process in a lecture at the Society of Arts:

[The naval architect] has to produce a structure possessing ample strength and stability, capable of meeting all conditions that may be encountered at sea, and of steaming at a desired speed over a stipulated distance. Safety, seaworthiness, and good behaviour must be ensured under conditions that cannot be determined with exactitude, as is the case with fixed engineering structures on land. … In addition the designer has laid upon him the individual requirements of naval officers, marine and mechanical engineers and artillerists: each class naturally thinking its department the most important.[48]

White did not assert a claim to full autonomy in the design process, nor did he claim – as Reed did – that the naval architect was a superior judge of ship design to the naval officer. He simply implied the claim, without drawing attention to it, that the naval architect was superior in the practice of ship design – something which had not been clear in the preceding fifty years. Perhaps more impressively, he did this without a great deal of controversy. Undoubtedly his educational work at the naval college, and the ways he nurtured credibility in his skills without controverting the authority of the officer, helped a great deal. Even Beresford, writing in his memoirs, singled out White's contribution to restoring the credibility and majesty of the Royal Navy's fleet.[49]

3. JACKY FISHER, PHILIP WATTS AND THE *DREADNOUGHT*

White was not the only person to benefit from the design meetings for the *Royal Sovereign*. When Fisher became First Sea Lord in 1904 he employed similar methods in the design of HMS *Dreadnought*. After a series of informal discussions about a big gun battleship, Fisher brought together a number of naval officers and civilians in the Admiralty, including White's successor as DNC, Philip Watts. Watts had entered the Royal School of Naval Architecture in 1866, then joined the Chief Constructor's staff, spending time in Torquay assisting William and Edmund Froude with test tank experiments on hull form. In 1885 he went to Armstrong's in the deal that brought White back to the Admiralty, before returning in 1902.

Watts held his predecessor at the Admiralty in high esteem. He also had great respect for Fisher. Writing with news of his appointment, he reminded the admiral of a conversation they had in the early 1880s:

I remember quite well when I was in the 'Inflexible' your promising to make me DNC when you became Controller, and thought of it more than once when White broke down. I suppose it may come yet, i.e. I may be working with you at the Admiralty, if you should come back in another capacity. I trust it may be so![50]

That opportunity came in 1904, when the two continued the close working relationship they had begun some twenty years previously. In Fisher's committee on designs, Watts played an important role conceiving the *Dreadnought*'s form, producing numerous designs and acting on the ideas of the committee members.

Fisher's design meetings differed from those that Hamilton held in 1888 in a number of ways. In structural terms, Fisher formed a committee that met on a regular basis over two months. Also, Fisher did not limit the membership of the group to service personnel, but invited a number of leading scientists and engineers, including the noted physicist William Thomson (Lord Kelvin) and the director of the Haslar test tank station, Edmund Froude. On a more personal level, other important distinctions can be identified in Fisher's particular approach. Fisher was an advocate of the view that naval power was best guaranteed by *matériel* strength, rather than a deep commitment to character and strategy informed by history.[51] To achieve that end he surrounded himself with a network of trusted technical specialists.

When Fisher was appointed Director of Naval Ordnance in 1886 he had already spent a number of years examining the *matériel* aspects of naval power. He had served on the gunnery staff of HMS *Excellent*, published *A short treatise on electricity and the management of electric torpedoes* (1868) and worked to establish the navy's first torpedo school. Fisher was undoubtedly an enthusiast for new technologies, eager to exploit them in the waging of war. But he did not work alone.[52] By the time Fisher was First Sea Lord he had built an extensive network of naval officers, engineers and scientific collaborators, the members of which tackled *matériel* problems and fed their insights back to the charismatic admiral. The design of the *Dreadnought* serves as a useful case study for analysing the issues of authority and expertise in this work. While the historiography has frequently sought to understand the *Dreadnought* as a product of Fisher's naval revolution, it can equally be argued that it was the product of committee design work and the negotiation of different types of authority.[53]

The idea of a big gun battleship capable of high speed was something that had long interested Fisher. As far back as January 1883 he had conceived of an 18-knot warship he called the *Nonsuch*.[54] Turning that idea into an actual battleship required a lot of work and expertise. Weeks after taking office as First Sea Lord, Fisher authored a document on 'The fighting characteristics of vessels of war' in which he provided a succinct description of his battleship philosophy, but also his approach to the design process. He was adamant that under his command, '*Strategy should govern the types of ships to be designed*'. This promised an extension of the rationale Hamilton had employed in 1888. Fisher was eager to ensure that the Admiralty did not return to older approaches to ship design, when politicians and admirals had been happy to follow the courses of modifying successful ship types and responding to work in foreign and private dockyards.[55] Fisher believed that it was much more important that Admiralty ship design should realise specific strategic principles and technological possibilities.

Fisher lost no time in directing W.H. Gard, assistant DNC, to design a 12-gun battleship. Fisher's professional relationship with Gard began in Malta, where he was Commander of the Mediterranean Fleet and Gard Chief Constructor. In 1902 he followed Fisher to Portsmouth, and then in 1904 to the Admiralty.[56] In December Gard wrote to Captain Charles Madden, another of Fisher's circle, about the design on which he was working: 'I think she will be liked – she has a lot of good points, and is as big as we can dock: *but not too big*. Anyhow, a battleship with her guns and at least 21 knots of speed must command respect if not admiration.'[57] Towards the end of 1905 Fisher formally brought many of the individuals with whom he had been discussing his ideas for battleship design together in a committee on designs.

The committee combined naval and civilian members, including Prince Louis of Battenberg (who was replaced by Captain C.L. Ottley when he took over as Director of Naval Intelligence), Engineer Vice-Admiral Sir Albert John Durston, Rear-Admiral Alfred L. Winsloe (commanding torpedo and submarine floatillas), Rear-Admiral Henry B. Jackson (Controller of the Navy), Captain Madden (Jackson's naval assistant), Captain John Jellicoe (DNO), Captain Reginald Bacon, Lord Kelvin, Philip Watts, John Harvard Biles (Glasgow professor of naval architecture), John Thornycroft (engine builder), Alexander Gracie (Fairfield ship design), R.E. Froude, Gard and Watts. The naval officers on the committee were largely all members of the 'Fishpond', who had gained vital experience working in departments linked to the production of naval technology.

Comparing the 1888 and 1905 groups reveals an important shift in the types of authority brought together to discuss ship design. Fisher had assembled a group with much stronger *matériel* and scientific interests. He also gave a strong indication in his opening remarks to the committee that the authority of science was, in some areas of ship design, non-negotiable. He was explicit that safety and stability were considered 'matters now made clear by scientific research, and require no further investigation'. The implication of Fisher's statement was that Watts would be the final arbiter of whether the design was safe. Having drawn the boundaries of expertise, Fisher authorised the naval officers present to use their 'experience' to 'propose the tactical and fighting requirements' for ships, and the civilian members to state 'the limits within which these requirements are capable of fulfilment'.[58] The present committee also differed from the working group that conceived of the *Royal Sovereign* both in its ambition and scope. It far outstripped that body in the detail with which it considered the design of new ships, its collaboration between naval officers and 'scientific, constructive, and engineering experts both in and outside the Admiralty' and its power to commission experiments and trials to aid its deliberations. Its members were informed they would form an 'advisory board of great value ... in the consideration of new types of vessels containing so many novel features'.[59] The committee was set certain fixed design particulars: 'Speed 21 knots. Armament 12-inch guns and anti-torpedo craft guns. ... No guns on the main deck, except anti-torpedo craft guns if necessary to place them there. Armour to be adequate. Docking facilities to be carefully observed.'[60]

The committee proceeded to discuss the design requirements Watts had to fulfil in his plans, while Watts began the laborious work of producing numerous designs. These designs were given alphabetical labels, and would be discussed and altered by the committee. The committee quickly settled on the inclusion of 12-inch guns, and Watts's first designs, E and F, featured 12 and 10 guns respectively. Neither design met with the committee's approval. There was a concern that the space between turrets was too close for the safe operation at the fore and aft stations. The close grouping of turrets in the mid ship was also deemed to afford the enemy 'an excellent "bulls-eye" at which to aim'. An alternative design was requested (G), with the expectation that the DNC would have 'sketch outlines ... ready for consideration at noon the following day'. In the subsequent meeting the committee examined a diagram showing the blast curves of various gun sizes, which seemed to demonstrate that the problems in design E still existed in design G. The decision was made to reposition a number of the fore and stern guns in a new design (D), which met with the committee's approval. The design also met with the approval of Sir Arthur Wilson and Beresford, who happened to be at the Admiralty when the committee met.[61]

The deliberations of Fisher's committee on designs suggest an important change in how naval officers and architects worked together. While polemicists like Reed had portrayed a fundamental gulf in the two groups' concerns, the work of the committee on design involved a great deal of collaboration. To address the issue of turret placement, naval and civilian members of the committee undertook experimental investigations and discussions. Captain Jellicoe oversaw a series of blast experiments conducted with seven different ships, from which he concluded that there must be 63 ft between the muzzle of any gun, port or sighting hood. Similarly, Captain Bacon and Professor Biles addressed the problem in written remarks for the committee's consideration.[62] The committee then concluded that a distance of 80 ft was required between each 12-inch gun. The DNC was then directed to 'embody this conclusion in the two designs' on which he was presently working.[63]

With the question of blast interference settled, the committee requested Watts to complete the details for design D and prepare a model 'with moveable turrets, conning towers, &c., in order that they might be able to appraise the extent of the interference by blast in the several dispositions of the guns corresponding to various possible arrangements of machinery, boilers, &c.'. After judging two variations of design D in model form, the committee requested a further design (H) that replaced two 'after broadside turrets' with one 'middle line turret'. Watts maintained his preference for design D, given that it was less disruptive to the division of the design's deck space (i.e. for stowing boats) than design H. The naval members unanimously disagreed, stating their refusal to 'sacrifice efficient gunfire to matters of less importance'.[64] By the terms Fisher had set the committee, this was a naval question. Having expressed the limitations, Watts now had to find a way to incorporate the wishes of the naval officers.

Similar discussions took place concerning the decision to fit the *Dreadnought* with Charles Parsons's marine steam turbine. Watts and Durston strongly urged the adoption of the engine, emphasising its 'simplicity' and 'saving of weight of about 1,000 tons [that] would be effected'. The naval members were less convinced, and requested 'fuller information as to the suitability of turbines for manoeuvring purposes'. Specifically, they doubted whether the comparatively small propellers that had to be used with a turbine system could bring a ship to rest as quickly as the larger propellers that were used in tandem with reciprocating engines. Again, a combination of experimental investigations and remarks from expert members of the committee provided the evidence on which the group was happy to proceed with the marine turbine. Fisher's committee approach to design provided a space in which very different types of experts working within the Royal Navy, and beyond, came into direct contact and openly shared suggestions. In this way the *Dreadnought* was a product of a design process that greatly differed from the authoritative approach of Admiralty boards settling design requirements themselves and extending patronage to a select few officials.

4. CONCLUSION

This article has proceeded with a central point to prove: experts had to gain authority. The power to act in ship design was contested, and ultimately granted to naval architects because they were able to nurture credibility and build important working relationships with officers and politicians. The opening overview of mid-nineteenth-century naval architecture and ship design revealed just how contested expertise was in this field. A number of groups worked to gain greater authority over ship design, including naval officers, shipwrights and naval architects. By highlighting a few flashpoints it is also possible to problematise authority and expertise within the setting of the Admiralty. From this it is possible to identify many different types of authority at play in this institution, including political, social, institutional, strategic and professional. We are left with a better understanding of the obstacles that naval architects navigated to gain authority, and the types of contests and institutional contexts in which their claims to expertise were made.

The subsequent case studies on the design meetings preceding the *Royal Sovereign* and *Dreadnought* reconstruct these contests and contexts. They allow us to chart precisely how naval architects navigated authority in the case of ship design. The 1888 and 1905 meetings highlight the importance of the institutional structures that White and Watts navigated, and the significance of having expertise acknowledged. From the parameters set by Hamilton and Fisher it is possible to mark the changing boundaries of authority and expertise in Admiralty ship design. But more importantly we follow the interactions between different experts and see how authority was negotiated. This was especially important in ship design, which necessarily involved a great deal of compromise. There was no such thing as the perfect ship, despite attempts in the 1860s to combine heavy armour, firepower and sail in unstable designs. The compromises at work in the design process were as much social as they were technical, given that ship design required the interaction of naval architects and officers, and the negotiation of their respective knowledge, skills and authority.

To gain authority legitimately in the Admiralty, one required a grant of power from a superior official. The professional and institutional activities that White and Watts undertook helped them to establish credibility at a time when naval architects practised an emergent expertise, and repeated ship design controversies drew the attention of a wide audience. Building credibility with naval officers and administrators could help their chances of gaining authority, but ultimately it came from being empowered to act. In their cases, the authority came from Hamilton and Fisher, and the guidelines they laid down for the design of the *Royal Sovereign* and *Dreadnought* respectively. Any history of expertise must keep those individuals who grant authority in the picture. Without them, the political, professional and institutional frameworks in which experts act will remain elusive

NOTES

I thank James Davey, Crosbie Smith, Larrie Ferreiro, Andrew Lambert and the participants of the 'Authority, expertise and the state, 1800–

1950' workshop at the National Maritime Museum for their insightful comments. I also thank archivists at Churchill College Cambridge.

1 [E.J. Reed], 'Admiralty programme of shipbuilding', *The Times*, 22 Mar. 1889, 13.

2 White, 'New battle ships', 194, 199.

3 Navy. Design for first class battle ships **1889** (C5635), 3.

4 White, 'New battle ships', 196.

5 *ibid.*

6 Leggett, 'Neptune's new clothes'.

7 White, 'New battle ships', 196–7.

8 *ibid.*, 184.

9 Reed, 'On the value of science to shipbuilders', 8–9.

10 Brown, *Warrior to Dreadnought*, 11, 26–7, 47–51.

11 Beeler, *Birth of the battleship*, 13–29, 87–107.

12 Lambert, *Last sailing battlefleet*; Parkes, *British battleships*; Baxter, *Introduction of the ironclad warship.*

13 Sharp, *Memoirs of William Symonds*, 38–92.

14 Lambert, *Last sailing battlefleet*, 27–34.

15 For Admiralty reform, see Hamilton, *Making the modern Admiralty*.

16 Debate on the appointment and qualifications of a Surveyor, quoted in [Chatfield], *Apology*, 6–7.

17 *ibid.*

18 *ibid.*, 5–6.

19 Baldwin Walker papers, D70/188 30(v) [cat. 64], Earl of Auckland to Baldwin Walker, 6 Aug. 1847.

20 Harris, *Our dockyards*, 36.

21 Russell, *Fleet of the future*, 13–16.

22 *ibid.*, 82.

23 [Reed], 'Introduction', xvi.

24 Leggett, *Shaping the Royal Navy*, chap. 3.

25 For a technical study of the ship's loss, see Brown, *Warrior to Dreadnought*, 47–54.

26 Leggett, *Shaping the Royal Navy*, chap. 4.

27 Russell, 'Loss of the "Captain" '.

28 An Old Sailor, 'Notes on the navy'.

29 Beeler, *Birth of the battleship*, 11.

30 Leggett, *Shaping the Royal Navy*, chap. 6.

31 William H. White to Captain Robinson, 16 May 1911 [a miscellaneous letter attached to a 1900 edition of White, *Manual of naval architecture* in the Caird Library at the National Maritime Museum].

32 Brassey, *British Navy*, vol. 4, 41.

33 [E.J. Reed], 'Dockyard administration', *The Times*, 5 Feb. 1877, 8.

34 Marsden and Smith, *Engineering empires*, 254–5.

35 Brown, *Century of naval construction*, 55, 60–1.

36 For details of the scheme, see Manning, *William White*, 89–93.

37 The full list consisted of George Hamilton, Arthur Hood, Anthony Hoskins, Admirals John Hopkins and Hotham, Arthur Forwood, William Dowell, Richard Vesey Hamilton, Frederick Richards, Vice-Admiral Baird, Walter Kerr, John Fisher, and William H. White.

38 Navy. Design for first class battle ships **1889** (C5635), 9.

39 White, 'New battle ships', 155.

40 *ibid.*

41 Brassey, *British Navy*, vol. 3, 32, 53; Manning, *William White*, 241.

42 White, 'New battle ships', 168–9.

43 White, 'Latest reconstruction of the navy', 541.

44 Navy. Design for first class battle ships **1889** (C5635), 9–11.

45 White, 'New battle ships', 162.

46 [E.J. Reed], 'Naval administration' *The Times*, 19 Jan. 1877, 6.

47 White, 'New battle ships', 181.

48 White, 'Modern warships: lecture I', 806.

49 Beresford, *Memoirs*, vol. 2, 351.

50 FISR 1/2 84, Philip Watts to John Fisher, 17 Dec. 1901.

51 For a fuller account of these schools of naval thought, see Semmel, *Liberalism and naval strategy*, 134–51. The term *matériel* was frequently used by actors in a way that today's commentators might talk about technology. For technology and actors' categories, see Marsden and Smith, *Engineering empires*, 3–4.

52 Smith, '*Dreadnought* science'.

53 Lambert, *Fisher's naval revolution*; Sumida, 'Fisher and the *Dreadnought*'.

54 On rereading his original note in 1910 Fisher added 'Nonsuch (of vast speed!)'. FISR 3/3/2025, Nathaniel Barnaby to Fisher, 15 Jan. 1910.

55 FISR 5/11/4218, Fisher, 'The fighting characteristics of vessels of war', 1904.

56 Andrew Lambert, 'Gard, William Henry (1854–1936)', *Oxford Dictionary of National Biography*, Oxford University Press, 2004.

57 FISR 5/11/4218, W.H. Gard to Charles Madden, 15 Dec. 1904.

58 Kemp, *Papers of Admiral Fisher*, vol. 1, 215–16.

59 *ibid.*, 200, 216.

60 *ibid.*, 201.

61 *ibid.*, 217–19, 223, 224, 225.

62 *ibid.*, 254–8.

63 *ibid.*, 233.

64 *ibid.*, 227, 229, 231.

REFERENCES

Primary sources: manuscripts

Churchill College Archive, University of Cambridge, John Fisher papers (FISR).

Jagger Library, University of Cape Town, Cape Town, Baldwin Walker papers.

CHAPTER 7

Ship impact model for technical assessment and selection of Carbon dioxide Reducing Technologies (CRTs)

J. Calleya,R.Pawling,A.Greig *

Marine Research Group, Department of Mechanical Engineering, University College London, Torrington Place, London WC1E 7JE, UK

ABSTRACT

It is not unreasonable to imagine that the future may herald higher energy prices and greater regulation of shipping's Greenhouse Gas (GHG) emissions. With the introduction of the Energy Efficiency Design Index (EEDI) into MARPOL Annex VI, tools are needed to assist Naval Architects and Marine Engineers to select the best solutions to meet evolving requirements for reduced fuel consumption and associated carbon dioxide emissions. To that end, a concept design tool, the Ship Impact Model (SIM), has been developed for quickly calculating the technical performance of a vessel with one or more Carbon dioxide Reducing Technologies (CRTs) at an early design stage. The underlying basis for this model is the calculation of changes from known 'baseline ships'. The Ship Impact Model has been used in two projects to assess which selection (individual or combination) of Carbon dioxide Reducing Technologies (CRTs) have the most potential, in terms of cost-effectiveness and under other technical, operational and regulatory influences.

KEYWORDS

Abatement measures; Carbon dioxide reducing measure; Carbon dioxide reducing technology; Decision support; Greenhouse gas; Ship design

1. BACKGROUND

Future Greenhouse Gas (GHG) emissions may be limited either due to regulation to reduce the risk of dangerous climate change or because the high prices of conventional GHG emitting energy products, such as fossil fuels, will lead to increases in energy efficiency and substitution by alternative energy sources. The credibility for such a future scenario is derived from energy scenarios, such as that derived by the International Energy Agency (International Energy Agency, 2010).

The efficiency increasing and energy options that are the most appealing are the options that reduce operating costs the most and have lower installation and purchase costs compared to conventional energy sources. Though it is important to consider that while an option can appear to be the most appealing on a basic cost and technical basis, as previous studies have shown (Eide et al., 2009 and Det Norske Veritas, 2010), the technical considerations are part of a more complex shipping system (Calleya et al., 2012) and take-up of new technologies can be conservative. Ship owners are nervous about investing in new technology. Long-term contracting arrangements can be complex (Pirrong, 1993); for instance a combined owner and operator may have more incentive to invest in energy efficiency (Rehmatulla, 2011). There is uncertainty about the performance of new technologies and, in particular, the future of fuel prices and legislation. For fuel prices, the concern is not just over the absolute price of fuel but also the differential between different fuel types. Invest too soon and they could be uncompetitive compared to owners who delay, invest too late and they may not meet the legislation deadline. Choose the wrong technology and it could be redundant in ten years time when further legislation on, for example, black carbon is introduced. The proposal at the 65th meeting of the IMO's Marine Environmental Protection Committee (MEPC 65) to delay the introduction of Tier III from 2016 to 2021 (IMO, 2013) is an illustration of why ship owners are reluctant to be early adopters of technology.

The purpose of the Ship Impact Model (SIM) is to provide a tool for ship owners and operators to permit them to quickly explore various scenarios with different combinations of Carbon dioxide Reducing Technologies (CRTs) in an uncertain future. The SIM was developed as part of the RCUK (EPSRC) funded project "Low Carbon Shipping – A Systems Approach" (LCS) (Smith, 2010) and then applied to a Rolls-Royce led project, the Energy Technology Institute Heavy Duty Vehicle Energy programme (ETI, 2013). Running from 2010 to 2013, the aim of LCS was to use understanding of the many components of the shipping system to explore how the shipping industry might respond to the challenge of a GHG constrained future from a combination of the technological, economic, logistical, operational and infrastructure perspectives. The scope of the overall project is described in Smith et al. (2010). This paper focuses on a specific research area within the project – the development of ship design and evaluation procedures for the estimation of the technical specification of new build and retrofit ships over the next 40 years.

2. DEFINITION OF CARBON DIOXIDE REDUCING MEASURES

Balland et al. (2010) work on air emission reduction methods, includes SO_x, and NO_x reducing measures in addition to Carbon dioxide (CO_2) and considers both operational and technological options as being air emission controls. The more generic term abatement options or abatement measures is often used by work submitted to the IMO's MEPC or published elsewhere (Buhaug et al., 2009, Det Norske Veritas and Lloyds Register, 2010 and Committee on Climate Change, 2011). Buhaug et al. (2009) makes a clear distinction between operational and technological measures, and operational measures have been assessed as likely to have an effect quickly and technical measures as likely to take longer to show a significant impact (Det Norske Veritas and Lloyds Register, 2010). However, there has been no clear definition of terms, possibly due to CO_2 reduction being associated with efficiency improvement. Although interlinked, there are key differences between reducing CO_2 emissions, reducing fuel consumption, reducing costs and increasing efficiency.

Two new specific terms, defined below, have arisen from the need for consistent use of language when modifying ships or ship operations with the prime objective of reducing Carbon dioxide (CO_2) emissions:

- A Carbon dioxide Reducing Measure (CRM) is any measure that reduces the Carbon dioxide (CO_2) emissions of a ship or a fleet of ships. A CRM can be categorised as an operational measure or a Carbon Dioxide Reducing Technology (CRT). An important operational measure (non-technological CRM) is reducing operational speed.

- A Carbon dioxide Reducing Technology (CRT) is any technology that can be incorporated into a ship (this could be either a retrofit or new build) that reduces the Carbon dioxide (CO_2) emissions of the ship compared to the original ship design (before modification). A CRT is one type of Carbon dioxide Reducing Measure (CRM) and can be categorised as reducing propulsion power, reducing auxiliary power, using fuel more efficiently (increasing energy/CO_2 emissions) or as using alternative fuels (e.g. Liquid Natural Gas (LNG) instead of Marine Diesel Oil (MDO)).

In this work the focus has been on CRTs that are currently available or are technically feasible today with little or no additional development cost. As ships have an expected service life of around 30 years, CRTs that are only applicable to new build ships will take much longer to widely implement (Calleya et al., 2012). Non-technological CRMs (such as trim optimisation and weather routing) do not interact with the ship in the same way as CRTs. Most non-technological CRMs are inherently cost-effective (often no physical changes to the ship are required) though the CO_2 reduction potential can be highly uncertain and may be dependent upon the incentive of the crew.

3. DEVELOPMENT AND DESCRIPTION OF THE SHIP IMPACT MODEL (SIM)

The SIM has been developed as a concept design tool for quickly calculating the technical performance of a vessel with different CRTs. It is intended for use at the early stage of a new build or retrofit design process. The underlying basis for this model is the calculation of changes from known baseline ships. Although there is a wide range of ship types in the global fleet that can be described in the SIM, the current work has focused on four ship types: container ships, bulk carriers, oil tankers and LNG tankers. These four cargo ships were chosen as they represent the largest proportion of shipping CO_2 emissions and account for 89% of global gross tonnage (Buhaug et al., 2009).

Table 1 summarises the selected ship designs that were generated. Surveys of the Clarksons Research Shipping Database, carried out at the start of the LCS project in 2010 and 2011, indicated that many ships fall into size categories of similar displacements, partly due to port and canal size limits. In the LCS project, the focus was on ship types where there is the biggest potential for CO_2 emissions reduction; ship size categories that carry the most overall deadweight and in which there are a lot of similar ships (in terms of speed and size). Note that Liquid Natural Gas (LNG) tankers are a relatively small proportion of current international shipping, but they were included as they may represent an emerging market with increasing use of natural gas for power generation. Two types of LNG tankers were examined, Heavy Fuel Oil (HFO) powered 2-stroke diesel propulsion and LNG powered steam turbine propulsion. In 2007 the majority of LNG tankers were steam turbine driven, after 2007 the trend has been towards more diesel powered ships (Noble, 2007). However, this means there are many steam turbine propulsion ships currently at sea (these variants are not shown in Table 1).

Table 1. Selected ship types, sizes and speeds: C=Container Ship B=Bulk Carrier O=Oil Tanker L=LNG Tanker.

Ship size categories	Displacement range (to nearest thousand tonnes)	10 knots	15 knots	20 knots	25 knots
Feeder container ship	14,000		C	C	C
Handysize bulk	33,000–34,000	B	B		
Handymax bulk	42,000	B	B		
Panamax ships	71,000–73,000	B O	C B O	C	C
Aframax oil tanker	117,000	O	O L	L	
Medium-sized LNG	115,000				
Post-panamax container ship	122,000–123,000		C	C	C
Suezmax tanker	147,000–165,000	O	O L	L	
Q-Flex tanker	154,000–155,000				
Ultra large container ship (ULCC)	172,000–175,000		C	C	C
Cape bulk	186,000–188,000	B	B		
Very large crude carrier (VLCC)	305,000–343,000	O	O		

The SIM uses a baseline data set of 36 full three-dimensional ship design models, these were produced using the ship design software Paramarine and are illustrated in Fig. 1. Each ship type was modelled at four sizes and two or three design speeds. These models were used to generate a virtual fleet of ships to populate the model. However, an operator or design house with their own fleet and design database could use this instead, so that the SIM could estimate the impact of CRTs on their vessels.

To include potential future changes in hullform design speed, ship models were made at one or two additional design speeds (depending on ship type). Slower (15 and 20 knot container ships and 10 knot oil tanker) hullforms were generated by manipulation of the extent of the parallel mid-body and bilge radius to achieve appropriate hull shape coefficients, such as prismatic coefficient and block coefficient. Hullforms with a larger block coefficient may have less CO_2 emissions per unit cargo carried due to an increase in cargo carrying capacity. This is in addition to ships operating at a lower Froude number (longer ships for the same speed), where viscous resistance is a much larger proportion of the overall resistance, having a lower overall resistance due to a hullform with a higher block coefficient. To find the lowest CO_2 emissions per unit cargo carried, over an operating profile, there may be a trade-off between cargo carried and the resistance of the hullform. This is similar to the Maersk Triple-E that has a fuller hullform (with a larger block coefficient) compared to the Emma Maersk Class (RINA, 2011) and is designed to operate at a number of different speeds, with an average operating speed of around 16 knots (in line with current slow-steaming practices; IMarEST, 2013).

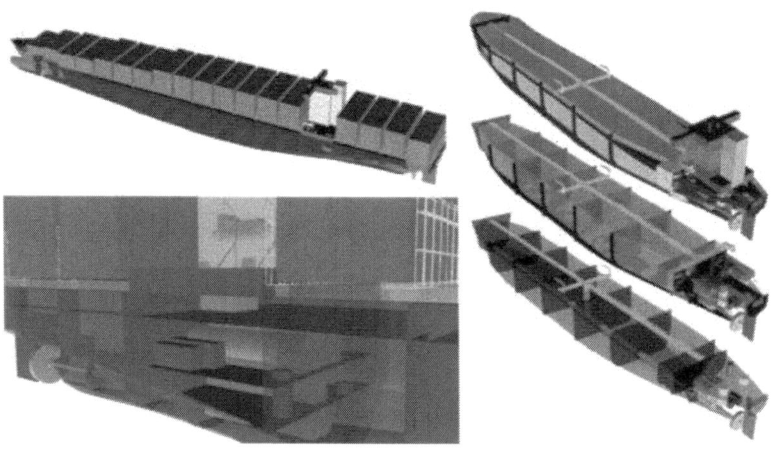

Figure 1. Paramarine models of a container ship (left), showing the detail of the engine room, and an oil tanker (right), showing the complete ship, cargo systems and propulsion machinery.

4. ASSUMPTIONS

Generally, most ships that carry cargo have an aft engine and machinery space and superstructure above this and holds along the length with ballast tanks around the cargo holds in varying forms (such as double bottom tanks, wing tanks, or hopper and topside tanks). The topological similarity of most ships within a given ship type allowed the simplification of the modelling approach. The impact of the CRTs was defined as only affecting certain parts of the ship, with other areas, such as crew accommodation, remaining unchanged. This permitted a much simpler and quicker iterative ship design spiral to be used to balance the design, compared to a more conventional design spiral where the entirety of the design might be changed in each iteration (Watson, 1998).

In the design process of the model (shown in Fig. 2), it was assumed that the engine was selected to operate at 75% of the engine MCR (a Maximum Continuous Rating of 75% this means that there is a combined sea and engine margin of 25%). The operational performance is evaluated over an operating profile, in the specific example that is mentioned here an operating profile described by Maersk was used (Cerup-Simonsen et al., 2009), and a 12% addition to the ship's viscous resistance was assumed to allow for hull fouling. This 12% fouling allowance is an approximation assuming 6 months between dockings using Brown's (1976) observations from warships of 1/8th per cent per day. Using 3 months (91.25) as the average fouling time period at a 0.125% increase per day gives a 12% increase in the overall resistance due to fouling and this was applied to the ship as an increase in viscous resistance. Applying fouling as a change in viscous resistance allows for the correct integration of CRTs, which can sometimes be a function of the resistance of the ship and may have a different effect on the performance of the ship depending on the level of fouling. In practice the fouling could vary a lot with a specific ship, likely as a function of the speed of the ship and sea conditions (such as water temperature).

The SIM uses Holtrop and Mennen (1982), updated to include those components that were necessary from Holtrop (1984). The propeller model uses polynomials that describe the Wageningen B-Screw Series of propellers (Oosterveld and Oossnan, 1975). A limited amount of comparisons were made with model test data, particularly with a chemical tanker model test report, provided by Marintek, in order to check the accuracy of the resistance and propeller models and to select which regression formulae to use.

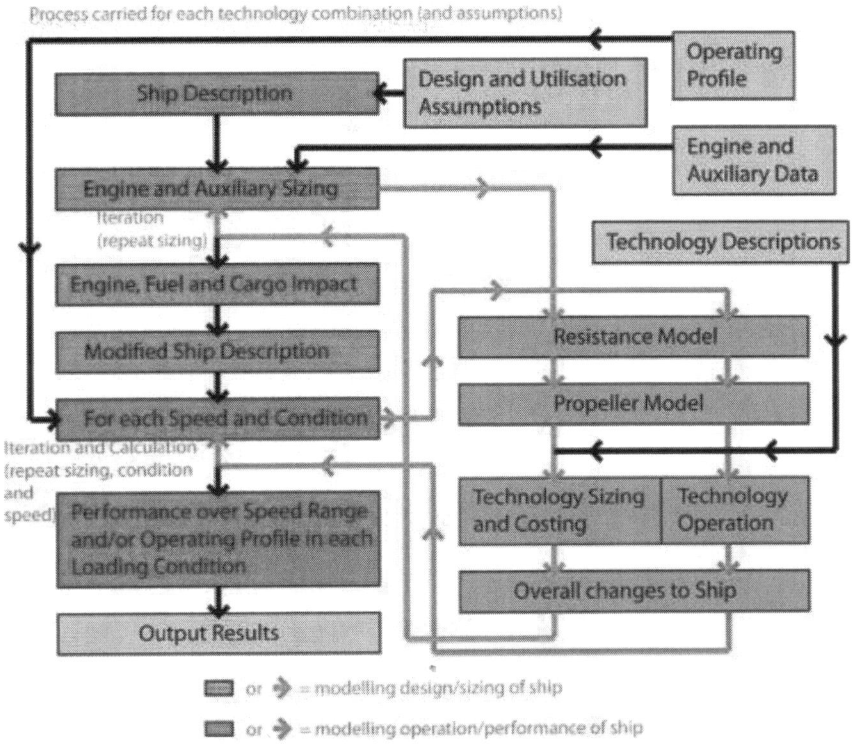

Figure 2. Ship Impact Model process flow diagram.

Another assumption that permits significant reductions in the computational time required is to assume that any mass changes to the ship impact on cargo capacity at a fixed displacement. This is true for primary (ship impact directly from the CRTs) or secondary (such as changing the installed engine size due to having a lower power requirement) mass changes. This approach has also been adopted in the Life Cycle Performance Tool (LCPA) developed by the EC funded project "Breakthrough in European Ship and Shipbuilding Technologies (BESST)" (BESST, 2013), where the ship impact is measured in terms of passenger cabins gained or lost. The alternative approach of maintaining cargo capacity and varying the displacement may require a much more sophisticated model. The hullform may have to change shape to reflect the change in displacement and resistance, requiring a highly parameterised model and an automated method of controlling it. This could be computationally intensive, requiring multiple iterative processes to generate the new hullform while assessing the impact of the change in resistance and propulsive machinery on the hullform.

Having every CRT impact on the rated cargo capacity could reduce the accuracy of a CRT's performance calculation as cargo ships are not always full to capacity (particularly for container ships and when in ballast for other ship types). In order to mitigate this, the operational design condition and operational ballast condition cargo capacities from the input ship are described separately from design capacity and are used for calculating operational performance.

The SIM works by assuming changes from a baseline ship and does not contain any stability or structural checks as this is not required for the fitting of the majority of CRTs provided the baseline ship already meets regulations. Any additional changes that are needed to meet ship safety and regulatory requirements are included in the technical assessment of the CRT. This means that specific stability and structural requirements, if required due to a CRT, such as sails, are represented in the file describing the CRT (described in Section 6) as a change in lightweight that is represented as an additional impact on cargo. Detailed three-dimensional models (as shown in Fig. 1) were used to create the input ship descriptions and were referred to in order to write the CRT files. For example, in order to correctly model the impact of adopting LNG, particularly the impact of the volume and the location of the LNG tanks, the detailed three-dimensional models that contained volume and stability calculations (as shown in Fig. 1) were needed in order to ensure that the simplified SIM correctly estimated the performance of LNG fuelled ships (Calleya et al., 2011). However, this does mean that if a large number of CRTs in combination cause a structural or stability change, this may not be modelled accurately.

Some assumptions contain a lot of uncertainty, particularly those relating to operational aspects, such as operational profile, auxiliary power utilisation (for example, the power requirement for refrigerated and chilled containers), displacement and cargo load when the ship is in operation (this is likely to vary between voyages). To evaluate the impact of uncertainty, the user can calculate the performance of the ship over up to two ranges of varying design and/or operational parameters (such as design speed, deadweight, design engine rating and additional resistance due to fouling). This means that the ship design process is automatically carried out again for each combination of design and/or operational parameters that are varied.

5. MODEL PROCESS OVERVIEW

Fig. 2 illustrates the process used in the SIM. This is carried out for the baseline ship and is then repeated for each technology or combination of technologies, as required. The first step is to size the baseline ship and select the installed power and support equipment. The SIM calculates secondary impacts, for example, if installing a CRT makes the engine more efficient, or reduces the energy requirement, then a smaller engine might be selected from the engine list. The SIM then carries out the same iterative process at each operational speed considering the off-design case where applicable. The propeller model is an example of this, for a fixed pitch propeller the pitch of the propeller is determined in the design stage and the same efficiency curve is used at all speeds to calculate the off-design performance of the propeller. This process is carried out both for the baseline ship and the ships with CRT(s) fitted. Modelling both the baseline and variants in the SIM – rather than comparing a SIM generated variant with a separate model – allows for a direct comparison of the results.

As mentioned in Section 4, it is possible to vary the inputs and assumptions, such as design speed, deadweight and endurance, repeating the analysis of the baseline ship and CRT(s) for each design speed, cargo capacity, endurance, fouling allowance or assumed auxiliary power requirement (power for crew and cargo). When a range of assumptions are explored that have an impact on the ship hullform parameters and cargo capacity (such as design speed or deadweight) the SIM estimates the new ship characteristics (including lightweight and hullform parameters) based upon curves fitted to the parameters of the available ships in the virtual fleet (described in Section 3). The type of curve fitter to each ship characteristic is fixed, and they were selected based on accuracy. This allows the model to represent a wide range of ships. Relationships such as that between wetted surface area, block coefficient, prismatic coefficient and displacement are not checked for consistency in the SIM and are instead derived from the input data. This avoids making additional assumptions, such as estimating wetted surface area, in the SIM and makes the SIM insensitive with regards to the source of the ship database.

As well as ship descriptions, technology descriptions and supporting data (engine list and associated fuels, boilers, description of fuels and LNG tanks, etc.) are also required. The correct supporting equipment is selected automatically depending on the technology description and allows secondary ship impacts to be calculated. For example, an LNG fuel demand will cause the required fuel to be calculated and LNG tanks sized to meet this capacity.

6. SELECTION AND DESCRIPTION OF CARBON DIOXIDE REDUCING TECHNOLOGIES (CRTS)

The technologies themselves are described as changes in up to 19 characteristics of the baseline ship. The changes to the baseline ship can be described in three categories, for different types of effect on the ship. The effects on the ship and how they are recorded are listed in Table 2.

The ship impact model technology interface must be flexible enough to deal with different technology architectures. To aid in the development of the interface, a reduced set of technologies was used to provide examples of the interface parameters required. The list in Table 2 was found to work well as an interface between the potential different CRT and ship combinations that could be considered. Most individual CRTs can be described by as few as three or four parameters from Table 2. This list was developed with input from subject matter experts on the CRTs. This was a two way process to ensure the right information is available from the ship and the ship represents the CRT in the correct way.

The CRT files (implemented as MATLAB functions) can access the following information about the ship and operating condition, as shown in Table 3. This means that the outputs from the CRT described in Table 2 can be described as functions of the inputs from the ship in Table 3.

Table 2. Possible inputs from CRT to Ship.

Type of parameter	Parameter	Units
Resistance parameters	Change in Viscous Resistance	%
	Change in Wetted Surface Area	± m^2
	Change in Propulsive Coefficient	%
	Change in Overall Resistance	± kN
Engine and fuel parameters	Change in Main Engine Specific Fuel Consumption	%
	Change in Auxiliary Engine Specific Fuel Consumption	%
	Change in Engine Shaft Speed	%
	Main Engine Fuel Selector	none
	Auxiliary Engine Fuel Selector	none
	Heat Boiler Fuel Selector	none
Service parameters	Change in main engine power	± kW
	Change in auxiliary engine power	± kW
	Change in shaft generator power	± kW
	Change in heat energy	± kW
	Deck space impact	± m^2
	Change in mass due to technology/cargo impact	± te
	Can technology be retrofitted?	Yes/No
Cost parameters	Unit purchase cost of technology	± $
	Through Life cost of technology	± $

A CRT file could be as simple as an engine improvement that reduces the specific fuel consumption by X%, this would not require any reference to the input list described above, in Table 3. A CRT file can also be a much more complex function of the inputs. Though a detailed customisable model could be described in a CRT file, this was only done in one instance. In some cases it was found to be more effective to look at a complex CRT, such as a waste heat recovery system, by precalculating a look-up table of different operational conditions that can be referred to by the SIM. The look-up table can be generated by the CRT file itself on the first iteration of the model or it can be generated in different software entirely and included inside the CRT file. This had the advantages that it was possible to interface with different software packages, such as the engine simulation software GT-Power for the ETI HDVE project, and that the run-time can be decreased by having pre-calculated results where full simulation is not required.

Table 3. Possible outputs from ship to CRT (if required).

Type of Parameter	Parameter	Units
Ship characteristics (51 items taken directly from full description of ship for a specific ship design speed and deadweight)	Cargo density	te/ m^3
	Available deck space	m^2
	Waterline length	m
	Beam	m
	Wetted surface area	m^2
	etc.	
Utilisation parameters (for particular speed and condition)	Operating speed	knots
	Demanded main engine power	kW
	Demanded auxiliary engine power	kW
	Demanded heat power (could be boiler power)	kW
	Demanded propeller torque	Nm
	Demanded retrofit?	Yes/ No
Design parameters (set by design condition)	Design speed	knots
	Installed main engine power	kW
	Installed auxiliary engine power	kW
	Installed shaft generator power	kW

The power from a waste heat recovery system is likely to be based on the specification of the host engine and could be described, in Eq. (1), as:

$$power_{waste\ heat} \propto \frac{OutputPower_{hostengine}}{RatedPower_{hostengine}} \quad (1)$$

There is also a choice as to whether the powerwaste heat is used for main propulsion power or hotel/auxiliary power. The waste heat recovery function is referred to each time the model looks at a different condition or ship speed to calculate powerwaste heat at the calculated Output Powerhost engine.

There is a lot of freedom to describe a range of CRTs. As shown in Table 3, the CRT can also be sized directly from the input ship description, which will not change between different ship operating conditions. For example, a Mewis duct, improving flow into a propeller, could be sized by referring to the propeller diameter in the ship description.

A separate fuel list is required to isolate fuel incompatibilities when looking at different CRT options and is needed to associate costs with the fuel(s) that the ship is using. Estimating cost is computationally more straightforward than calculating performance. The cost is important in deciding which technologies to pick over others and, in the case where a CRT can be implemented to a varying extent, how much of a CRT should be used, such as when considering using different numbers of standardised rigid sails.

7. OUTPUT FORMAT AND UTILISATION OF THE SHIP IMPACT MODEL

The ship impact model has had two main objectives; to advise, as part of the LCS project (Calleya et al., 2012 and Smith et al.,), on what carbon dioxide emission reductions are possible for shipping and how best to regulate shipping to achieve this; and to act as an early stage design tool, as part of the Rolls-Royce led ETI HDVE project, to select which technologies to develop further.

The initial technical output from the ship impact model is saved to a spreadsheet in the format depicted in Fig. 3. This output format was developed to allow post processing of the SIM outputs to provide a simplified description of the ship impacts of CRTs that could be used in a model of the wider shipping system without requiring run-time execution of the SIM to generate each of thousands of ships in the global fleet, this is described in Calleya et al. (Calleya et al., 2012). This output structure also allows subject matter experts to assess the performance and impact of their CRT on the ship.

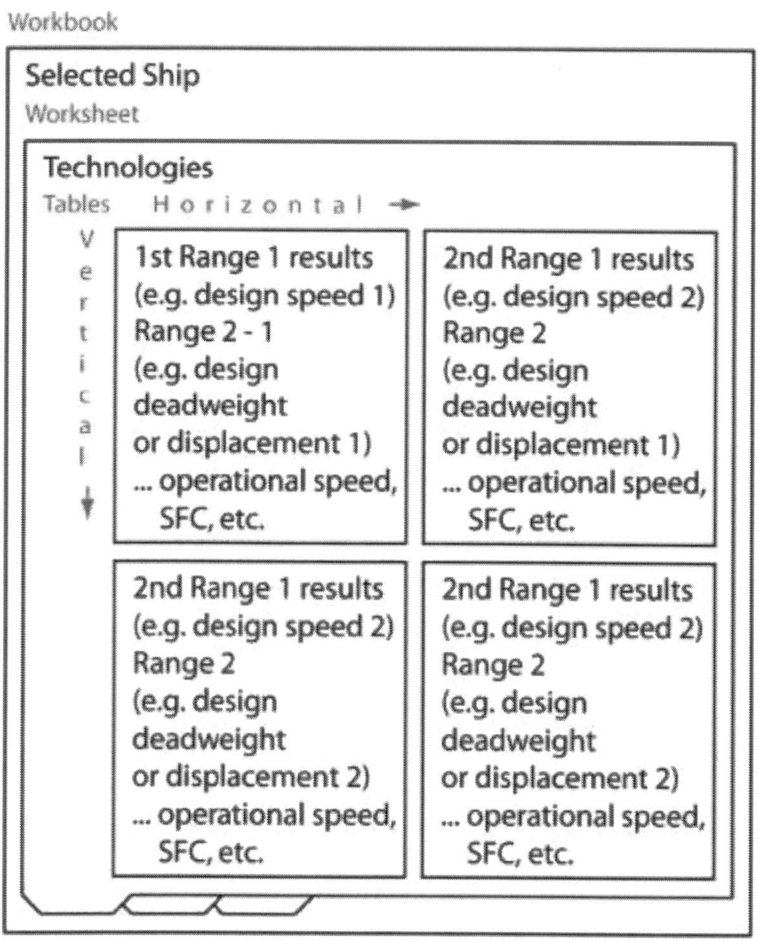

Figure 3. Format of the Ship Impact Model (SIM) output spreadsheet.

8. MULTIPLE CARBON DIOXIDE REDUCING TECHNOLOGIES (CRTS)

Each individual technology is described using the interface as described in Section 6. The same characteristics, shown in Table 2, are used to combine the effects of different CRTs together. At the level of detail shown in Table 2 most of the individual characteristics of each CRT can be added or multiplied together, the overall effect of this on the whole ship is much more complex.

In order to demonstrate the SIM using combinations of CRTs four CRTs were selected. These were; Propeller Boss Cap Fin (PBCF), changing fuel to Liquid Natural Gas (LNG), photovoltaic solar cells for auxiliary power, and thrust from wing sails. The effect of the individual CRTs and combinations are shown in Table 4 and Fig. 4. The ship selected for this analysis was a 25 knot Panamax Container ship, with a design deadweight of 35032 t (4584 TEU) and 294.2 m in length, with an endurance of 35 days and an operating profile described by Maersk (Cerup-Simonsen et al., 2009), as mentioned in Section 4. The time in design condition and ballast condition was assumed to be 68% and 12%, respectively, the remaining 20% of time was in port. The CRTs themselves were picked from independent studies done for both LCS (Calleya et al., 2012 and Smith et al.,) and ETI projects, with the help of subject matter experts.

In the specific example, shown in Table 4, solar cells and wing sails achieve only a small beneficial effect due to the described ship having a very small amount of available deck space and a high design speed. The wing sail assumed a north Atlantic crossing, with wind speeds around force four, based on a north Atlantic pilot chart (Gibbons-Neff and Miller, 2011). Wind power to provide thrust can have a much more significant impact at lower speeds and in ships with more space and stability available to fit them, such as on a bulk carrier. These differences between ship types are captured in both the CRT files and the ship descriptions, which give the amount of available deck space for fitting CRTs that have to go on deck.

Table 4. Some outputs for a 25 knot Panamax container ship with an endurance of 35 days.

CRT	Baseline	PBCF	solar cells	LNG	wing sails	PBCF, solar cells, LNG and wing sails
CO_2 (tonnes/day)	272.4	266.6	272.4	237.6	269.0	229.9
% reduction	0.0%	2.1%	< 0.1%	12.8%	1.2%	15.6%
CO_2/design cargo	0.0078	0.0076	0.0078	0.0097	0.0077	0.0094
% reduction	0.0%	2.1%	< 0.1%	−24.8%	1.2%	−20.8%
EEDI (gCO_2/t nm)	43.0	42.7	43.0	53.4	42.9	53.0
% reduction	0.0%	0.6%	< 0.1%	−24.2%	0.1%	−23.3%

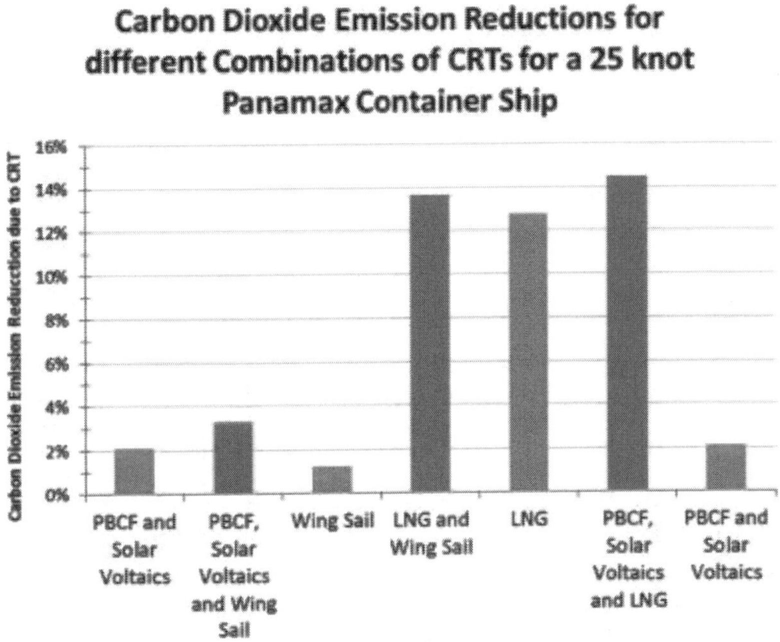

Figure 4. CO_2 emission reduction of different CRTs. Combinations of CRTs are shown in between single CRTs (with the exception of PBCF and Solar Voltaics).

It has been assumed that all CRT and ship combinations are designed to the same endurance of 35 days, as large cargo ships tend to have a large bunker capacity. For the purpose of this study the same endurance was used for LNG as for HFO fuelled ships.

The final column of Table 4 shows the effect of combining all four CRTs on a single ship, a total of 15.6% reduction in CO_2 emissions is achieved. This clearly demonstrates that simple addition of the individual contributions from CRTs (this gives 16.1%) over estimates the net reduction from a suite of CRTs. A better estimate is found by calculating the combined effect of a suite of CRTs by representing individual CRTs (*A*, *B* and *C*) in terms of the percentage CO_2 reduction (CO_2%reduction) as:

$$CO_2\%\text{reduction}_{A+B+\dots} = 1-(1-CO_2\%\text{reduction}_A) \times (1-CO_2\%\text{reduction}_B)\times \dots \quad (2)$$

Using Equation (2) produces a value of 15.7%, this still does not give exactly the same result as that calculated by the SIM, over estimating the net reduction from a suite of CRTs.

9. DISCUSSION

Published CO_2 and fuel savings, particularly from CRT manufacturers, can be given in isolation of their ship application context. Performance estimates can vary greatly depending on the assumptions made and the performance measure that is being used.

Table 4 shows that for LNG while the overall CO_2 emissions of the ship have decreased, a reduction in cargo capacity has meant that the CO_2 emissions per unit cargo has increased. The large increase in CO_2 emissions per unit cargo is due to the cargo space taken up by using a less dense fuel (compared to HFO/MDO) that has to be contained in pressurised tanks while maintaining the large bunker capacity used for a fast (25 knot) Panamax Container Ship with an endurance of 35 days. Table 4 also shows that this results in a reduced EEDI compared to the baseline ship despite using LNG that has a lower carbon factor than HFO/MDO. It may be necessary to re-consider the large bunker capacities of existing ships when switching to less dense fuels, especially fuels that require additional containment, compared to HFO/MDO.

Table 4 also shows that the percentage reduction in terms of the Energy Efficiency Design Index (EEDI) is much higher than the percentage reduction in CO_2 emissions.

The Ship Impact Model allows design decisions to be made based on an operating profile, though accurately considering the operational characteristics can be difficult due to the lack of available operational data and its uncertainty.

10. LIMITATIONS OF THE SHIP IMPACT MODEL (SIM)

Some CRTs, such as sails, are effected by environmental conditions (wind and waves). Even when not considering the sea conditions, it is unclear whether hydrodynamic and propulsor related CRTs can be scaled accurately (to different ship types, sizes and speeds) and it may not be possible to calculate some combinations of hydrodynamic and propulsor related CRTs from independent performance information because they interact with each other directly (for example, flow control devices before and after a propeller) and have to be modelled or tested together.

Only three ship loading conditions were considered; a design condition, a ballast condition and a port condition. It may be necessary to consider more operating conditions to better calculate the CO_2 emissions of a ship or to examine the potential change in CO_2 emissions due to varying design and operational assumptions.

This paper has focused on the CO_2 emissions from CRTs, however it is the task of the ship designer to evaluate the options considering both cost and operational performance; there are two different ways of doing this; net present value and required Freight Rate (Stopford, 2009).

11. FUTURE WORK

Future work should aim to better understand some of the uncertainties in the modelling of CRMs, particularly those aspects that are difficult to model, such as environmental and operational conditions. In particular, more emphasis is needed on:

- The affect of different design and operational assumptions on CO_2 emissions. A sensitivity analysis can be used to explore changes in the design and operational assumptions.
- Operational states, in terms of energy usage. More operational states should be included to allow for a better estimate of the fuel consumption. In particular, more hull fouling conditions should be examined because hull fouling may have a large affect on the

resistance of a ship and a high degree of uncertainty. Considering some aspects of the ship in the time domain, such as hull fouling or energy storage, may allow a better estimation of the performance of a ship and CRMs.

- Alternative modelling methods to increase flexibility and accuracy. For example, having an alternative regression method for resistance that can be used instead of Holtrop-Mennen.
- Modelling and designing systems to represent synergistic combinations of CRMs.

Designing for operational flexibility considering both operational data and the environment may allow some ships to operate with lower CO_2 emissions and may allow a better estimate of sea and engine margins.

It is also necessary to consider net present value or required freight rate and regulation (particularly on SO_x and NO_x emissions) as incentives to the adoption of CRMs.

There may also be the potential to explore the trade-off between cargo capacity and resistance in terms of the CO_2 emissions per cargo carried by changing the hull shape.

ACKNOWLEDGEMENTS

The main body of this work was developed as part of the RCUK Energy funded research project "Low Carbon Shipping – A Systems Approach", Grant reference is EP/H020012/1.

The assistance from Marintek, Rolls Royce and the Energy Technologies Institute, as part of "Heavy Duty Vehicle Efficiency – Marine", provided invaluable information to develop and validate the model, and is gratefully acknowledged. Especially Marintek for being open and accessible about the information they could provide.

REFERENCES

1. Balland, O., Solem, S., Hagen, A., Erikstad, S., 2010. A decision support framework for the concurrent selection of multiple air emission controls. In: Proceedings of the 9th International Conference on Computer and Information Technology Applications in the Maritime Industries.

2. BESST Breakthrough in European Ship and Shipbuilding Technologies, 2013. (accessed on October 2013) ⟨http://www.besst.it/BESST/index.xhtml⟩.

3. Brown, D.K., 1976. Fouling and economic ship performance. In: Proceedings of the 5th Inter-Naval Corrosion Conference.

4. Buhaug, Ø.; Corbett, J.; Endresen, Ø.; Eyring, V.; Faber, J.; Hanayama, S.; Lee, D.; Lee, D.; Lindstad, H.; Markowska, A.; Mjelde, A.; Nelissen, D.; Nilsen, J.; Pålsson, C.; Winebrake, J.; Wu, W.-Q., Yoshida, K., 2009. Second IMO GHG study 2009 (MEPC 59/INF.10), Marine Environment Protection Committee – 59th Session, International Maritime Organisation.

5. Calleya, J., Pawling, R., Greig, A., Bucknall, R., 2011. Assessing the Dependence of Natural Gas on the Size and Topology of Container Carriers and Other Ship Types. Gas Fuelled Ships 2011, Motorship, Rotterdam.

6. Calleya, J., Pawling, R., Smith, T.W.P. , Greig, A., 2012. Ship design and evaluation for a ghg constrained future. The Environmentally Friendly Ship, RINA, London, 28–29 February 2012, (shorter re-print under the title 'Calibrating the Future' The Naval Architect, RINA, in January 2013).

7. Cerup-Simonsen, B., de Kat, J., Jakobsen, O., Pedersen, L., Petersen, J., Posborg, T., 2009. An integrated approach towards cost-effective operation of ships with reduced GHG emissions. Trans. – Soc. Nav. Arch. Mar. Eng. vol. 117, 1–18.

8. Committee on Climate Change, 2011. Review of UK Shipping Emissions, (accessed in July 2013) ⟨http://www.theccc.org.uk/publication/review-of-uk-shipping-emissions⟩.

9. Det Norske Veritas and Lloyds Register, 2010. Assessment of IMO Energy Efficiency Measures (MEPC60/INF.18). Marine Environment Protection Committee – 60th session.

10. Det Norske Veritas, 2010. Pathways to low carbon shipping – Abatement potential towards 2030. ⟨http://www.dnv.com/resources/reports/pathwaystolowcarbon shipping.asp⟩, (accessed in July 2011).

11. Eide, M.S., Endresen, O., Skjong, R., Longva, T., Alvik, S., 2009. Cost-effectiveness assessment of CO2 reducing measures in shipping. Maritime Policy Manag. vol. 36, 367–384.

12. Energy Technology Institute (ETI), 2013. Universities and Science Minister David Willetts visits d2m ETI project involving Rolls-Royce & UCL, Energy Technolo- gies Institute website, Available from: ⟨http://www.eti.co.uk/news/article/uni versities_and_science_minister_david_willetts_visits_2m_eti_project_invo⟩, (accessed in October 2013).

13. Gibbons-Neff, P. and Miller, P., 2011. Route planning for a micro-transat voyage. In: Proceedings of the 4th International Robotic Sailing Conference, United States Naval Academy, USA, Lubeck, Germany, 16–17 August, Table 2, ⟨http://www.usna.edu/Users/naome/phmiller/IRSC%20USNA%20routing%20paper.pdf⟩,(accessed in March 2013).

14. Holtrop, J., Mennen, G., 1982. An approximate power prediction method. Int. Shipbuild. Prog. 29, 166–170.

15. Holtrop, J., 1984. A Statistical Re-Analysis of Resistance and Propulsion Data International Shipbuilding Progress 31, 272–276.

16. IMarEST, 2013. Maersk Mc-Kinney Moller, Shipping World & Shipbuilder, pp.18–23.

17. IMO Press Briefing, 2013. IMO Marine Environment Protection Committee 65th session pushes forward with energy-efficiency implementation ⟨http://www. imo.org/MediaCentre/PressBriefings/Pages/18-MEPC65ENDS.aspx⟩, (accessed in August 2013).

18. International Energy Agency, 2010. World Energy Outlook 2010, ⟨http://www. worldenergyoutlook.org/media/weo2010.pdf⟩, (accessed December 2013).

19. Noble, P.G., 2007. The next generation of LNG carriers for long distance and harsh environments, Intertanko, Houston, ⟨http://www.ivt.ntnu.no/ept/fag/tep4215/innhold/LNG%20Conferences/2007/fscommand/PS6_3_Noble_s.pdf⟩, (accessed in July 2013).

20. Oosterveld, M., Oossnan, P., 1975. Further computer-analyzed data of the wagenin- gen B-screw series. Int. Shipbuild. Prog. vol. 22, 2–14.

21. Pirrong, S.C., 1993. Contracting practices in bulk shipping markets: a transactions cost explanation. J. Law Econ. vol. 36, 937–976 (accessed in February 2013).

22. Rehmatulla, N., 2011. Applying systems thinking approach for qualitative analysis of GHG emissions regulations in shipping. In: Proceedings of the 1st International Conference on Marine and Maritime Affairs, Plymouth.

23. RINA, 2011. Expensive, elaborate and exclusive Triple-E takes centre stage, The Naval Architect, 30.

24. Smith, T.W.P., Bucknall, R.W.G., Dinwoodie, J., Gibbs, D., Mangan,D.J., Turan,O., 2010. Low carbon shipping – a systems approach, ship design and operation for environmental sustainability. In: Proceedings of the RINA.

25. Stopford, M., 2009. Maritime Economics. Routledge. Watson, D.G.M., 1998. Practical Ship Design. Elsevier Science Ltd., pp. 49–50.

CHAPTER 8

Delhi's Pacific Ambition: Naval Power, "Look East," and India's Emerging Influence in the Asia-Pacific

Walter C. Ladwig III

ABSTRACT

While India may not traditionally be considered to be a significant actor in the Asia-Pacific region, over the past 18 years, New Delhi has undertaken a concerted effort to direct its foreign, economic, and military policies eastward. What began as economic cooperation with the nations of Southeast Asia has expanded into full-spectrum engagement with the major powers of East Asia. This article explores India's emergence in the Asia-Pacific, concluding that, while in the near term India's presence and influence will be felt most strongly in Southeast Asia, a steadily expanding economy, paired with a growing partnership with key regional actors and an increasingly capable Navy, positions the South Asian giant to have an impact on the emerging security architecture of the region.

When considering the major powers of the Asia-Pacific region, India may not come readily to mind. Indeed there is a sharp division of scholarly opinion on the role that the South Asian giant plays there. In a recent academic text on the politics of the region, Derek McDougall dismisses India in a single sentence as a country that "interacts with the Asia-Pacific in various ways." [1] In contrast, in an article published the very same year, Harsh Pant firmly asserted that India is already "a major player in the Asia-Pacific regional balance of power along with the U.S., China and Japan." [2] As with many things in life, the truth lies somewhere between these two extremes. While it is premature to describe India as a "major player" in the Asia-Pacific, it is also a mistake to ignore the impact that its emerging role can have in shaping the future of the region.

Over the past 18 years, New Delhi has undertaken a concerted effort to direct its foreign, economic, and military policies eastward. What began as economic cooperation with the nations of Southeast Asia has expanded into full-spectrum engagement with the major powers of East Asia, such as Japan and the United States. India's expanding role in the Asia-Pacific has been facilitated by countries such as Singapore, Vietnam and Indonesia that look to the South Asian giant to help hedge against Beijing's growing regional influence. A steadily expanding economy, paired with a growing partnership with key regional actors and an increasingly capable Navy, positions India to have an impact on the emerging security architecture of the Asia-Pacific.

This article explores India's emergence in the Asia-Pacific region in seven parts. First, India's complex relations with China are discussed as they influence aspects of India's eastward orientation. This is followed by a discussion of India's emerging naval power and its view of its area of interest in Asia. Discussion of India's eastward orientation begins with Southeast Asia before moving on to East Asia, Australia, and the United States. Finally, after exploring several potential constraints on India's ability to act as an extra-regional power, the paper concludes with a discussion of the impact India can have on the future of the regional order in the Asia-Pacific.

1. THE CHINA FACTOR IN INDIA'S MOVE TO THE EAST

India's emerging influence in East and Southeast Asia is driven by a host of factors including geography, economics, and historical ties. Nevertheless, in understanding India's push toward the Pacific Ocean, it is difficult to ignore the role played by its northern neighbor. India's relations with China are complicated. On the one hand, economic cooperation and enhanced political ties benefit both nations: bilateral trade between the two Asian giants stands at nearly $25 billion per year, and both sides have pledged to double that amount by 2010. [3] China has recently displaced the United States as India's largest trading partner while India is China's ninth-largest market. [4] On the political front, having fully normalized relations in 2003, Beijing entered into what it calls a "Strategic and Cooperative Partnership for Peace and Prosperity" with New Delhi in 2005. [5] This latter development signals an important upgrade in relations between the two countries, and a sign that Chinese officials recognize India will become a significant Asian power. [6] The two nations share a desire to see the international sphere transition to a multi-polar structure in which each country has an increased voice in global affairs. Military relations between the two neighbors have also steadily improved, with an agreement in 2006 to begin undertaking joint military exercises, as well as high-level exchanges between their respective armed forces. [7]

Balancing these positive developments, however, is long-standing friction: their 1962 war inflicted a humiliating defeat on India and created an unresolved border dispute. [8] Such security tensions are not helped by China's on-going military buildup. Furthermore, China has been a principal supplier of weapons technology, both conventional and nuclear, to Pakistan, India's South Asian bête noire. In July 2007, Beijing deepened this "all-weather friendship" when it signed a free trade agreement with Islamabad. On the political front, India desires to be recognized as a great power in the international order. It is jealous of the status accorded to China by its seat on the United Nations Security Council and its recognition as an official nuclear power under the Nuclear Non-Proliferation Treaty.

The potential for discord between the two countries can be clearly seen in the energy sector. Beijing is desperate to secure hydrocarbon resources for its expanding economy, while India is increasingly reliant on similar energy sources. In recent years, China has beaten India in head-to-head competition for oil assets in Kazakhstan, Ecuador, and Nigeria. [9] Despite the pledges of future cooperation in pursuit of energy supplies, Šumit Ganguly has argued that "India is in a fundamentally competitive if not conflictual relationship with China" in their joint quest for energy resources. [10]

China's efforts to secure its access to overseas energy resources have brought it into India's backyard. [11] Oil from East Africa and the Persian Gulf must cross the Indian Ocean to make its way to the market in China. In an effort to secure its interests, China has helped establish a network of ports and partnerships with countries in the littoral region – including several nations that have traditionally been hostile to India. Indian observers frequently suggest that the goal of this so-called "string of pearls" strategy is to secure access to locations that could be used to project Chinese power into the Indian Ocean. Regardless of whether or not the "string of pearls" is an accurate characterization of Beijing's Indian Ocean strategy, China has certainly been active in the region. [12] To the west of India, China financed the construction of a major port complex for Pakistan at Gwadar, which some analysts presume will be used by the People's Liberation Army Navy (PLAN) to gain a strategic position in the Arabian Sea, close to the mouth of the Persian Gulf. [13] To the east, the Chinese military has reportedly assisted Burma with the construction of several naval facilities on the Bay of Bengal – particularly at Kyaukpyu and Hainggyi Island. [14] As with Gwadar, there has been speculation that these facilities are being upgraded to serve China's needs in a future military contingency. To the south, China recently reached an agreement to develop a port project for Sri Lanka at Hambantota on the island's south coast. [15] China's support for Pakistan, as well as its encroachment into the Indian Ocean, is viewed by some as part of a coherent strategy to encircle India and confine its influence to South Asia. [16] These concerns are accompanied by apprehension over the PLAN's on-going expansion, which is viewed as a possible threat to India's strategic interests in the region. [17] Not surprisingly, India's Foreign Minister recently described the rise of China as one of India's foremost security challenges. [18]

Notwithstanding Chinese rhetoric about a desire for a peaceful rise, among Indian policymakers there is a general view that China represents a long-term economic, and possibly military, competitor to India. [19] While China's long-term ambitions are far from clear, scholars such as Ashok Kapur perceive that Beijing seeks to achieve hegemony in Asia. [20] However, disagreement has developed in the Indian foreign policy establishment over how to respond to the challenge posed by China. Pragmatists, as Mohan Malik calls them, believe that economic engagement and wary cooperation can occur between the two countries. [21] While India does need to take steps to guard against Chinese power, cooperation should be the first priority. In line with India's traditional preferences, pragmatists favor internal balancing – a self-reliant buildup of military power. On the other hand, the so-called hyper-realists view China as an immediate threat to India. They argue that increased economic engagement has not led to a reduction in China's support for Pakistan or its efforts to encircle India. As a result, they advocate an aggressive strategy of alliance-building with other states in Asia to constrain Beijing's ability to dominate the region. A third group, the so-called appeasers, believe that China is a fundamentally peaceful state that does not pose a threat to India.

The composition of these groups is not fixed, but, in general, elements of the Congress Party and India's business community, as well as the Ministries of Finance and Commerce, incline towards the pragmatist view; segments of the BJP party and the Ministries of Defense and Home Affairs fall into the hyper-realist camp; while India's vocal Communist Party of India (Marxist) and other left-wing parties tend to ascribe to the appeaser view. In the face of this division of elite opinion, the Indian public has expressed an increasing wariness of China. A 2008 Pew attitude survey found that a plurality of Indians believe that China's economic growth is bad for India, while a super-majority views China's increasing military power negatively. [22] This marks a noticeable deterioration in Indian perceptions of China compared to just three years ago. [23]

While pragmatists shape India's policy towards China, at present it is not clear that economic interaction will necessarily foster a constructive relationship between the two countries. Several prominent scholars of Sino-Indian relations believe that interaction between Asia's two rising powers is more likely to be characterized by geopolitical rivalry and competition than cooperation. [24] From the Chinese perspective, Hong Kong-based scholar Chung Chien Peng contends that China's and India's mutual aspirations to great power status will necessarily produce a competitive relationship in Asia – a view shared by Chinese analysts who perceive rivalry to be at the core of Sino-Indian relations. [25]

As India's eastward focus demonstrates, Delhi's engagement with China is coupled with efforts to lay the groundwork for a more robust strategy should this pragmatic approach fail to deliver results. Indian leaders frequently state that they are not seeking to contain China, but their policies indicate that they are hedging their bets. India's efforts to expand its presence in the Asia-Pacific can be seen as part of a hedging strategy that develops economic linkages and security cooperation with key states in the region wary of Beijing's power, while still maintaining mutually beneficial economic ties with China.

2. INDIA'S GROWING MARITIME POWER AND STRATEGIC VISION

In the time of the Raj, British India managed the Empire "from the Swahili coasts to the Persian Gulf and eastwards to the Straits of Malacca." [26] Security of British commerce in the Far East and the trade routes to Australia and New Zealand depended on India's power. [27] British India's geographic position, military strength, and natural resources led the Viceroy, Lord Curzon, to proclaim that the master of India would become the greatest power on the Asian continent. [28]

As India achieved independence, British officials assumed that the Raj's extra-regional power would naturally pass to the Republic of India. [29] While India was unable to play such a role in the decades after independence, it finally seemed ready to seize the inheritance of the Raj in the late 1990s when the right-of-center BJP government launched an ambitious program of naval acquisition that was paired with a "forward-leaning" foreign policy that sought to cement India's access and political leverage across the littoral region from East Africa to the Asia-Pacific. [30] In 2000, then-Defense Minister George Fernandes declared that India's maritime "area of interest … extends from the North of the Arabian Sea to the South China Sea." [31] Although the Indian Navy's 2004 maritime doctrine points to a narrower domain, arguing that "for the first quarter of the 21st century [India] must look at the arc from the Persian Gulf to the Straits of Malacca as a legitimate area of interest," the actions of the Indian Navy, as discussed below, suggest that the true scope of Delhi's ambition is closer to Fernandes' formulation. [32] In identifying this expansive sphere of influence, Indian strategists are staking an explicit claim to the legacy of the British Empire as the natural boundaries of India's influence. [33]

Given this expansive zone of interest, it is not surprising that some scholars expect that an emerging India will seek to become an extra-regional power in key sub-regions of the Asia-Pacific. [34] These suppositions appeared to be confirmed in 2004, when Prime Minister Manmohan Singh asserted that India's "strategic footprint covers … South East Asia and beyond," a geopolitical scope that "should inform and animate our strategic thinking and defense planning." [35] Echoing the view of India's civilian leadership, the then-Chief of Naval Staff (CNS), Admiral Arun Prakash, argued in 2005 that "it is imperative for India … to retain a strong maritime capability in order to maintain a balance of maritime power in the Indian Ocean, as well as the larger Asia-Pacific region." [36] The present CNS, Admiral Sureesh Mehta, appears to share his predecessor's views, as he recently suggested that "[the Navy's area of] interest is not restricted to the Indian Ocean." [37]

It is also possible to divine the scope of a state's ambitions by what it does, in addition to what it says. If the Indian Navy were primarily concerned with Pakistan or littoral defense, then a localized fleet consisting of short-range surface combatants supplemented by land-based naval aviation assets would be most appropriate. However, since a nearly decade-long procurement holiday came to an end in the mid-1990s the world's fifth largest navy has instead undertaken a sustained buildup that seeks to develop power projection capabilities, particularly the capacity to conduct expeditionary warfare. Admiral Mehta recently outlined an ambitious goal: "by 2022, we plan to have 160-plus ship navy, including three aircraft carriers, 60 major combatants, including submarines, and close to 400 aircraft of different types." [38] In support of those efforts, the Navy's share of India's expanding defense budget has steadily risen from 11.2 percent in 1992–93 to 18.3 percent in 2007–08. [39] The primary question for the Indian Navy's ambitious modernization program is whether or not it will succeed in achieving its goals in the specified timeframe. Although the pattern of the Navy's ambition is clear – both foreign acquisitions and indigenous construction projects focus on long-range operations and extra-regional power projection – India's naval acquisition and indigenous construction projects have been notorious for their delays.

The Navy's ability to deploy and sustain its aircraft carrier is the key to projecting its maritime power abroad. Naval planners envision a three-carrier fleet as the cornerstone of India's future blue-water navy, which would allow India to maintain at least one at sea at all times, and would make it the equal of the Royal Navy, and second only to the United States, in aircraft carrier assets. However, realization of this goal may be slow in developing. After extensive life-extension programs, India's present carrier, the 45-year-old, 28,000-ton *INS Viraat*, is expected to hang on until 2013 when the Navy hopes to finally be able to take delivery of the long-delayed Russian-built *Admiral Gorshkov*. [40] This 44,500-ton capital ship, which will be renamed the *INS Vikramaditya*, will carry 16 fighter aircraft and another half-a-dozen anti-submarine warfare helicopters. With a range of 14,000 nautical miles, the *Vikramaditya* will have three times the operational radius of the *Viraat*, which the Navy's leadership believes will provide a “quantum jump” for its maritime capabilities. [41] However, should the *Gorshokov* be further delayed until 2015, as some informed observers speculate, the Navy could find itself without a carrier for several years. [42]

Supplementing the *Vikramaditya* will be India's first indigenously constructed aircraft carrier. The induction of this 40,000 ton platform, which has a complement of 16 fighter aircraft and 20 helicopters, has been officially delayed until 2015, however 2018 may be a more realistic date. [43] Construction of a second indigenous carrier, expected to commence some time after 2010, could provide a three-carrier fleet by the mid-2020s, allowing a carrier to be assigned to three different areas – the Indian Ocean, the Arabian Sea and the Bay of Bengal. [44]

The increased range of the new carriers, along with the upgrade in attack aircraft from the old Mk. 51 Sea Harriers to the supersonic MIG-29K

Fulcrums, will not only increase India's force projection capacity by an order of magnitude, but it will provide India with the most powerful strike aircraft in the region – a significant development as the Indian Navy is keen to boost its ability to decisively influence military operations on land. [45] The acquisition of advanced airborne command-and-control aircraft, as well as tankers capable of in-flight refueling, further facilitates the sustained projection of Indian airpower.

While carriers excel at power projection, carrying out sea control and sea denial missions falls to the submarine fleet. As Barry Posen notes, a submarine fleet is a key asset of anti-submarine warfare capability, "which in turn is the key to maintaining command of the sea." [46] This is a particular issue for the Navy as its existing submarine force of ten Russian-built Kilo-class boats and four German Type-1500s has only a 50 percent operational readiness rate. [47] India's submarine force received a further setback in late 2008 when the nuclear-powered Akula-II class attack submarine that the Navy was widely believed to have leased from Russia suffered a fatal accident during sea-trials, leading Moscow to back out of the deal. [48] Nevertheless, the submarine fleet is being upgraded to enhance its littoral warfare capabilities by introducing the Russian-developed *Klub* naval cruise missile, which is capable of striking land and naval targets from a range of 275km. For the first time, Indian submarines will have the ability to launch missiles at surface targets while submerged. After technology transfer issues delayed the start of construction for over a year, six French-designed *Scorpene* diesel-electric submarines, capable of launching both *Klub* and *Exocet* anti-ship missiles, will be entering service between 2013 and 2017. [49] A tender for six additional diesel-electric submarines, featuring air-independent propulsion which will greatly increase their operational range, is expected to be made in 2009, however this would merely replace the submarines expected to be decommissioned in the next six years. [50] A contract with Boeing for eight P-8 long-range maritime reconnaissance aircraft will enhance the Navy's anti-submarine warfare capabilities – particularly if the Navy exercises the option to purchase an additional eight planes. [51]

The Indian Navy's strike capability received a major boost in 2006 with the introduction of the supersonic *BraMos* multi-role cruise missile. [52] A joint venture between Russia and India, the 290-km range *BraMos* can receive guidance and targeting information in flight from Unmanned Aerial Vehicles (UAVs) and patrol aircraft. Multiple variants under development will allow the missile to be launched by aircraft, surface ships, and submarines.

At present, large portions of India's fleet of 57 surface combatants, including five of its eight destroyers, seven of its eleven frigates, and many patrol craft are approaching the end of their service life. While the Navy's current procurement plans will keep the absolute size of the surface fleet static, replacing these obsolete vessels with modern versions will noticeably enhance the capabilities of the fleet. As of mid-2007, the Navy had 33 ships under construction that were expected to enter service in the next five years. [53] This includes three new destroyers and eight new frigates that are capable of employing both the *Klub* and *BraMos* missiles. India is presently in negotiations with Russia for the purchase of several additional *BraMos*-capable Talwar-class stealth frigates. [54]

India's ability to conduct amphibious operations beyond its territorial waters was enhanced by its June 2007 acquisition of a 16,900 ton Landing Platform Dock (LPD), from the US in its first-ever purchase of an American ship. Capable of transporting nearly 1,000 soldiers and six medium-sized helicopters, the *Jalashva* will address the Navy's lack of heavy sea-lift and mass-landing capability, while also providing the ability to function as a command-and-control platform for fleet operations in an amphibious landing or emergency response scenario. The Navy is already considering acquiring a second LPD from the US, while its present complement of *Magar*-class Landing Ship Tanks, which can transport 500 men or 15 armored vehicles, will be supplemented by four new amphibious assault ships. [55] The Navy's complement of three fleet replenishment tankers, which buttress extra-regional operations, will receive an addition in 2010. With several more tankers and the recently announced plans to acquire eight new minesweepers, India would be on the verge of possessing Asia's only viable expeditionary naval force.

While Indian naval enthusiasts can rightly say that their Navy is in the process of developing a reach that extends across "the entire maritime swath from [the] Western Pacific Ocean through the Straits of Malacca into the Indian Ocean," there is still a considerable distance to go before that ambition is a reality. [56] The ability to conduct sustained expeditionary operations requires a fleet to possess the ability to defend its sea lanes, counter mines, defend against air attack and project firepower on land. The Indian Navy is attempting to address these issues, but in recent years Russia has proven to be a less than reliable supplier of naval platforms while indigenous programs suffer from endless delays. Although the Navy is unlikely to reach its ambitious goals before the mid-2020s, as discussed in the following section, it already possesses the ability to conduct meaningful operations beyond the Indian Ocean. [57]

3. INDIA LOOKS EAST …

With the end of the Cold War and subsequent collapse of the Soviet Union, India not only lost its primary patron, it also lost its main trading partner, arms supplier, and source of subsidized oil. At the same time, the end of the bipolar struggle between the superpowers freed Asia from many of the ideological divisions that had defined it in previous decades. Desiring a way to link India's own liberalizing economy to those of Southeast Asia, as well as cultivate a greater role in the region, Prime Minister P. V. Rao launched the "Look East" policy in 1991. This multi-pronged initiative sought to create strategic political and economic ties with individual nations in Southeast Asia while simultaneously developing closer ties with the Association of Southeast Asian Nations (ASEAN), the region's premiere multilateral economic and political institution. Rather than being simply an economic policy, Look East marked "a strategic shift in India's vision of the world and India's place in the evolving global economy." [58]

Over the past 16 years, India has steadily expanded and strengthened its relationship with ASEAN. It first became a sectoral dialogue partner in 1992, engaging with member states on issues such as trade, investment, and tourism. As a result of the "growing ties between the two sides that have developed in recent years," India became a full dialogue partner in 1995. [59] The following year, India began to participate in the ASEAN Regional Forum (ARF) – a regional security grouping that included Asia-Pacific powers such as Australia, China, Japan, and the US as well as the EU and Russia. This marked a dramatic reversal of Indian policy, which had previously been strongly opposed to participating in multilateral security fora. In 2002, the relationship was upgraded again when the first ASEAN–India summit was held. The following year, India became one of the first non-Southeast Asian nations to accede to ASEAN's Treaty of Amity and Cooperation, which commits India to upholding the provisions of ASEAN's 1967 charter – including adhering to the principles of non-aggression and non-interference in the internal affairs of partner nations. [60] In combination with the ARF, this step significantly bolstered India–ASEAN security cooperation.

With the policy supported by both BJP and Congress governments, Look East and its associated strategies to establish India's presence in Asia have become an institutionalized component of the country's foreign policy. This approach has met with success because it achieves important foreign policy goals for both India and its partners. Increased engagement in the region is part of New Delhi's overall effort to increase its presence in an area where its sphere of influence overlaps with that of Beijing. [61] For ASEAN member countries, India provides an alternative that allows them to reduce their economic dependence on both China and Japan. [62] Unlike those two East Asian powers, India does not have a historical legacy of invasion or domination in the region. Deeper integration with India allows ASEAN nations to tap into a dynamic liberalizing economy in a democratic state. Not surprisingly, Singapore's Foreign Minister has noted that, "We see India's presence as being a beneficial and beneficent one to all of us in South-east Asia." [63]

India's economic engagement with the region has expanded by an order of magnitude since 1990 as its annual trade with ASEAN nations grew from $2.4 billion to over $30 billion by 2007, with a goal of expanding bilateral trade to $50 billion by 2010. As a result of these increasing ties, India has reached an agreement with ASEAN to create a free trade zone by 2012 that would link 1.6 billion people in an area with a combined GDP of over $1.5 trillion. These measures are welcomed in the region, because they allow countries to avoid economic dependence on a single market. As Singapore's deputy Prime Minister has argued, "For Southeast Asia, a dynamic India would counterbalance the pull of the Chinese economy, and offer a more diversified basis for prosperity." [64] Complementing its economic and political linkages to Southeast Asia, India has taken steps to achieve physical linkage as well. These efforts include the construction of a rail link between Hanoi and New Delhi that passes through Burma, Thailand, Laos, and Cambodia, as well as a major highway project linking India to Thailand via Burma. [65]

4. NAVAL DIPLOMACY

As Amitav Acharya argues, the ASEAN countries "recognize India's role as a possible counterweight to any potential future threat from China." [66] Lee Kuan Yew, Southeast Asia's elder statesman, has described India as "a useful balance to China's heft" in the region, while Malaysia's Foreign Minister has discreetly suggested that the South Asian power was "strategically located to provide [Southeast Asian nations] the necessary umbrella as a big country." [67] More bluntly, veteran Singaporean diplomat K. Kesavapany has noted that ASEAN countries "envisage India as acting as a counter-balance to a possibly overdominant China in the future," while Meidyatama Suryodiningrat, the editor of the Jakarta Post, has called on India to "become a psychological deterrent to China's increasing influence and gradual domination of this region." [68]

As a result, India's engagement with Southeast Asia has not been limited to economics. Military exercises have been a key means by which India has asserted its presence in the region. Since 1991, India has periodically held joint naval exercises with Singapore, Malaysia, and Indonesia in the Indian Ocean. In subsequent years, it has undertaken bilateral exercises with Vietnam, Thailand, and the Philippines. In 1995, this military engagement matured into the annual *Milan* series of naval maneuvers that India conducts with ASEAN nations in the Bay of Bengal. Not only do such exercises showcase India's naval capabilities but they also contribute to enhanced interoperability with regional navies and can positively shape perceptions of shared security concerns.

India has also dispatched its vessels on forward presence missions designed to "show the flag" in the South China Sea and beyond. In support of such operations, Indian ships, including the aircraft carrier *INS Viraat* have made high-profile port calls in major cities such as Manila, Jakarta, Singapore and Saigon. [69] This activity has been followed up in subsequent years by bilateral exercises in the South China Sea with the navies of Singapore, Vietnam, and the Philippines. [70] In undertaking such missions, India demonstrates the blue-water capabilities of its Navy in a maritime domain that China has previously claimed exclusively as its own. [71]

To facilitate power projection into the Asia-Pacific, the Indian Navy is upgrading its base network. A second naval base on India's eastern shore is being constructed near Vizag, 30 miles south of the existing Eastern Naval Command HQ. This new facility, which is expected to house two aircraft carriers as well as attack submarines and support ships, is reportedly intended to facilitate "ongoing Indian naval exercises in the South China Sea with the navies of China-wary countries." [72] The Indian Navy has also announced plans to bolster its forces deployed in the east, which officials connect to India's broader eastward focus, noting that "the naval fleet in east India has long legs and, with the government's emphasis on the 'Look East' policy, we are strengthening the east now." [73]

In 2005 a Far Eastern Naval Command (FENC) was established at Port Blair in the Andaman Islands, which are located midway between the Bay of Bengal and the Straits of Malacca – a key chokepoint linking the Indian Ocean to the South China Sea. By 2012 it is expected that the FENC will consist of three main bases and a network of anchor stations that will house surface combatants and patrol vessels as well as submarines. By some accounts, the completed facility will be larger than the former US naval facility at Subic Bay in the Philippines, which was America's largest overseas naval base during the Cold War. [74] Significantly, airfields in the Andamans bring the Straits, as well as large portions of the South China Sea, within the operational radius of India's frontline fighter aircraft, such as the SU-30K Sukhoi and SEPECAT Jaguar. [75] While notionally intended to facilitate control over the eastern straits that are vital to the trade routes of the Indian Ocean, some believe that expansion of the FENC is intended to check China's influence in the region. [76] In the words of Raja Mohan, the Navy's new eastward orientation enables India "to be a significant player in the emerging Asian balance of power." [77]

The Indian Navy's engagement with Southeast Asia is not simply about power projection; it has also attempted to cultivate soft power within the region. India's reputation as a positive force in the region was enhanced in the wake of the 2004 tsunami when the Indian Navy undertook its largest peacetime mission ever to provide assistance to the peoples of Indonesia, Thailand, and Malaysia. Thirty-two ships and over 20,000 naval personnel assisted the host nation governments in locating survivors and evacuating casualties, as well as providing emergency sources of power and water. [78] Following the Indian Navy's high-profile role in escorting US military supply ships and other high-value vessels through the Straits after the September 11, 2001 attacks, India has attempted to demonstrate its ability to provide regional "public goods" by offering to assist in securing key sea lanes in Southeast Asia. [79] So far its offers to assist littoral nations in patrolling the Straits of Malacca have been rebuffed, however the Indian Navy has begun to conduct coordinated anti-piracy exercises in the northern approaches to the Straits with both the Indonesian Navy and the Royal Thai Navy.

5. BILATERAL RELATIONS

Through its engagement, in the form of increased trade and military cooperation, India enhances the ability of Southeast Asian nations to internally balance against domination by any single power. India's closest regional ties are with Singapore, which has been a strong proponent of India's engagement with ASEAN. Singapore was instrumental in securing India's association with the ARF and has publicly supported India's bid for a seat on the UN Security Council. Regular naval exercises between the Royal Singapore Navy and the Indian Navy, SIMBEX, have been held since 1993. These strong Indo-Singaporean defense ties led to a 2003 defense cooperation agreement that made the city-state India's most important bilateral partner in the region. Then Indian CNS Admiral Prakash suggested that India's defense ties with Singapore are "possibly the closest that we have ever been to any country." [80] The closeness of these links between the two countries can be seen in the fact that personnel from the Singaporean Army, Navy, and Air Force all train at facilities in India and weapons systems for their fleet are tested at India's Chandipur firing range. This cooperation with India facilitates Singapore's own security strategy, which is based on "borrowing political and military strength" from countries outside of Southeast Asia. [81]

India also has a long history of cooperation with Vietnam, having supported its intervention in Cambodia in 1979. Since that time, New Delhi has extended Hanoi numerous investment lines of credit worth nearly $110 million, and bilateral trade with Southeast Asia's fastest growing economy reached $3 billion in 2008. [82] In recent years, the rise of China has highlighted shared strategic concerns between the two countries: both states share a land border with China, with which they have each fought wars and have outstanding territorial disputes. [83] A 1994 agreement on India–Vietnam defense cooperation was followed by an additional agreement in 2000 that established regular discussions between the two countries' defense ministers and set the stage for joint naval exercises. In July 2007, Vietnam and India agreed to "diversify and deepen" their relationship by expanding trade, enhancing scientific cooperation, and undertaking collaboration on civilian nuclear energy, as well as seeking to "strengthen cooperation in defense supplies, joint projects, training cooperation, and intelligence exchanges." [84] There is speculation in some circles that India seeks to secure access to port facilities at Cam Ranh Bay, the finest natural harbor in Asia, which could be used to "balance" the Chinese presence in the Indian Ocean. [85]

Historically, Indonesia has supported India's enhanced engagement with ASEAN, and as Jakarta emerges from a decade of turmoil, it has the potential to be an important regional partner. Indian officials recognize Indonesia as "the largest and most influential member of ASEAN," whose economy is expected to rival that of major European countries in aggregate size by 2020. [86] In 2005, India and Indonesia agreed to establish a strategic partnership based on "shared values and commitment to democracy ..." [87] This agreement both deepened and broadened political, economic, and security ties. A follow-on accord opened the possibility of jointly producing military hardware. [88] Aside from the military benefits that would accrue from jointly procuring and developing defense material, Indonesian analysts note that "working with India would be a way for Indonesia to help ASEAN nations check the power of China in the region." [89] Other regional observers concur, noting that India–Indonesia security agreements look "like an attempt to constrain Beijing's run as the region's top dog." [90]

Although not as robust as its links to Singapore, Vietnam, and Indonesia, India has also enhanced its economic and security ties with Malaysia, Thailand, and the Philippines as a part of Look East. India is Malaysia's largest trading partner in South Asia and the two nations have established numerous joint ventures in fields ranging from transportation to information technology. India has provided training for the Malaysian Navy's fighter pilots, submarine personnel and special forces; its dry docks have undertaken the repair and refit of several Malaysian naval vessels; and the two navies have undertaken joint exercises. [91] As a concrete sign of these expanding ties, India has raised the idea of selling *BraMos* missiles to Malaysia, an offer that would only be made to countries that were "considered as strategic partners in India's security equation." [92] For its part, the Malaysian Foreign Minister has called for a "strategic alliance" with India. [93] India established a framework agreement for a free trade area with Thailand in 2003 and has signed numerous agreements on economic cooperation in sectors ranging from agriculture to tourism. In the political realm, Thailand has shared Singapore's interest in encouraging India's involvement in Southeast Asia: former Thai Prime Minister Thaksin Shinawatra made three official visits to India between 2001 and 2005, while interim Prime Minister General Surayud Chulanont made a visit in 2007. On the security front, India and Thailand have entered into agreements to share terrorism-related intelligence and tactics. [94] A 2005 memorandum of understanding between the Indian Navy and the Royal Thai Navy established procedures for coordinated maritime patrols. Following a state visit by then-President Kalam to Manila in 2006, India signed a defense agreement with the Philippines that would deepen maritime cooperation and allow bilateral military exchanges. [95] Indo-Philippine ties are relatively immature at present but can be expected to grow in the future.

India's economic and military engagement with Southeast Asia is perceived to be a tangible manifestation of its strategic intention to compete with China for influence in the region. For example, Arthur Waldron argues that India's move into the region "is a good example of how China's buildup is already eliciting counterbalancing responses around her periphery." [96] Similarly, Amitav Acharya notes that "there is considerable evidence" that India is balancing China in Southeast Asia, while Zhao Hong believes that New Delhi and Beijing will engage in a "continuing and long-term competition for regional influence in the ASEAN region." [97] Some security analysts have suggested that, "by extending its area of operations firmly into the South China Sea, India presents a direct challenge to China." [98] With the Indian Navy active in the South China Sea, the PLAN's hegemony over its own backyard has been called into question. As Raja Mohan and Parag Khanna argue, India's efforts send a message that it "will not simply cede primacy [in Southeast Asia] to China." [99] Such a competing influence can be beneficial for the nations of Southeast Asia, which have historically had difficulty preserving their autonomy in the presence of great powers. However, without a legacy of dominating the region, India does not provoke this same anxiety. [100] India's increasing economic and military links broaden the range of powers influencing Southeast Asia, which allows local states to adopt hedging strategies versus China – developing ties with New Delhi while maintaining relations with Beijing. [101]

6. … ON TO NORTHEAST ASIA …

The scope of India's ambition as an Asian power is not confined to Southeast Asia. After its initial success with ASEAN, India has moved into phase two of its Look East policy, which encompasses a region "extending from Australia to East Asia." [102] Indian officials envision playing "an ever increasing role" in this "extended neighborhood." Simultaneously India is expanding the range of issues on which it engages East Asian nations from trade to wider economic and security issues, representing a further "strategic shift in India's vision." [103]

Despite Chinese efforts to curtail its influence, India gained political acceptance in its bid to be recognized as an Asia-Pacific power in 2005 when it was invited to attend the inaugural East Asian Summit – an effort some believed would be the stepping stone to the formation of an "East Asian Community" to mirror the European Community. [104] Support for India's inclusion in the EAS came from Southeast Asian nations such as Singapore, Indonesia, and Thailand, as well as Japan and South Korea – all of which championed India's participation despite objections from China. [105] While some Indian commentators view their nation's inclusion in Asia-Pacific regional fora as "a recognition of [India's] fast growing economic and political clout," analysts taking a realistic view of events in Asia recognize that India was not invited to attend the EAS based on its economy alone, but also to prevent Beijing from dominating the institution. [106]

As with Southeast Asia, India has paired political ties with defense diplomacy to enhance its presence in the region. [107] India conducted joint naval maneuvers with the South Korean Navy in 2000, 2004, and again in 2006. [108] Although often overlooked, the South Korean Navy possesses a sizeable complement of surface combatants and submarines, comparable to the navies of France and the United Kingdom. May 2007 marked the first ever visit by a South Korean Defense Minister to India. This was coupled with expanded political ties as New Delhi and Seoul established a "long-term cooperative partnership for peace and prosperity" that is intended to take Indo-Korean relations to "a higher level." The framework included economic cooperation and expanded trade ties, as well as a foreign policy and security dialogue that promotes bilateral defense cooperation. [109]

Seoul is particularly concerned that China's on-going military buildup will enable it to dominate the sea lanes of the South China Sea – a development that would significantly undercut Seoul's political independence from its giant neighbor. As a result, Seoul has actively supported India's naval presence in maritime Asia to offset China's regional power. Despite Chinese opposition, South Korea has championed India's inclusion in East Asian regional fora. [110] New Delhi and Seoul are also united in their concerns about the proliferation of nuclear weapons and missile technology in their respective sub-regions. These worries converge in China which has aided both Pakistan and North Korea with their nuclear weapons programs. [111] Subsequent cooperation between Islamabad and Pyongyang in a "nukes-for-missiles barter trade" reinforces the perception that India and South Korea face a common challenge. [112]

India has fashioned an even stronger strategic partnership with Japan. Unlike many countries in Asia, India bears no historical animus towards the Japanese. Tokyo and New Delhi's shared interests in restraining the scope of China's influence in Asia, as well as their "deep interest in tackling regional and global security challenges," have led to a strengthening of defense ties that one overly exuberant South Asian commentator has termed an "Asia-Pacific alliance between India and Japan." [113] Although it has been increasingly common to focus on China as the leading power in East Asia, it should not be forgotten that Japan's economy is roughly the size of China's and India's combined and with a defense budget that exceeds $40 billion, its military is among the most advanced in the world. [114] In particular, Japan's Maritime Self-Defense Force is easily the most capable indigenous Navy in the Asia Pacific, which "will likely continue to 'outclass' those of regional rivals for the foreseeable future, in spite of recent modernization efforts within the Chinese navy and air forces." [115]

A host of factors are driving enhanced cooperation between India and Japan. On a geopolitical level, they can both be considered potential rivals to China for primacy in the broader region. As Japan continues to evolve into a normal nation willing to undertake a regional military role, tensions – both historic and strategic – continue to plague its relations with China. The military buildup undertaken by Beijing in the past decade has concerned both Japan and India, particularly because some experts predict that by early next decade, China's military could overtake Japan's as the foremost military force in Northeast Asia. [116] In an effort to forestall competition from its southern and eastern neighbors, China has attempted to prevent both Japan and India from gaining equal international status by opposing expansion of the UN Security Council to include the two nations, resisting the legitimization of India's nuclear arsenal, and attempting to block India's participation in pan-Asian regional fora. [117] Such clumsy efforts have only had the effect of driving New Delhi and Tokyo closer together.

This is not to suggest that ties between India and Japan are driven strictly by realist geopolitical considerations. Among the rising powers of Asia, both Japan and India are established democracies while China remains an autocratic state. [118] As an editorial in Japan's largest daily newspaper argued "India is an extremely important partner with which Japan can shape a new international order in East Asia because the two countries share common values of freedom and democracy." [119] Former Japanese Prime Minister Shinzo Abe had emphasized the importance of institutionalizing liberal values such as human rights, the rule of law, and democracy in Asia. This focus dovetailed nicely with enhanced ties with the world's largest democracy. The notion of relying on shared principles to support strategic dialogue took a concrete step in May 2007, when, at a meeting of the ASEAN Regional Forum, senior leaders from Japan and India joined their counterparts from the US and Australia for consultations among the "democratic quad" in Asia. [120]

Following an agreement to strengthen cooperation between the two navies, India and Japan conducted reciprocal naval exercises in the Indian Ocean and the Sea of Japan in 2005. [121] The following year, the service chiefs of all three branches of the Japanese Self-Defense Forces made official visits to India, while the Indian Defense Minister, Pranab Mukherjee, visited Tokyo for consultations with his counterparts, which produced an agreement to promote defense exchanges between the two countries. During Prime Minster Singh's visit to Japan in December 2006, the two countries established a framework to transform their relationship into a strategic partnership that would impact all aspects of interstate ties from trade and investment to defense cooperation. [122] This was followed by a 2008 Joint Declaration on Security Cooperation between Japan and India that the two nations claim will form an "essential pillar for the future architecture" of security in Asia. [123] This marks only the second such security agreement that Japan has ever entered into.

India is already the leading recipient of Japanese overseas development aid, and Tokyo has supported major infrastructure projects within India. [124] Commenting on the significance of enhanced Indo-Japanese ties, then-Prime Minster Abe suggested that this will become Japan's "most important bilateral relationship in the world." [125] Given the importance of Japan's security alliance with the United States, this is a bold pronouncement. India appears to be putting similar weight on the bilateral relationship: As former Indian Foreign Secretary Lalit Mansingh has noted, "if we are forced to choose between China and Japan, my bet will be on Japan." [126] A number of Japanese and Indian scholars assess that the intensifying strategic partnership between Delhi and Tokyo is part of a concerted effort to build an Asian regional order that counters China's increasing power. [127] As India's naval capabilities mature in coming years, its expanding security ties with Japan and South Korea are positioning the South Asian giant to have a significant influence on the multipolar maritime balance of power that already exists in the region between Japan, China and the United States – an outcome that would support Delhi's desire to see a "dynamic, multi-polar Asia." [128]

Despite the great public enthusiasm, there are reasons to be somewhat more circumspect when examining Indo-Japanese ties. Economic engagement between the two countries has failed to keep pace with the development of security ties: while there is strong evidence that Japanese companies are increasingly viewing India as an alternative to China for foreign direct investment, trade between the two nations is quite low, with Sino-Japanese trade totaling more than twenty-times that of Indo-Japanese trade. [129] Moreover, some critics contend that the much-hyped 2008 joint declaration does little to move Indo-Japanese ties beyond their prior 2006 agreement. [130] Nevertheless, given the negligible diplomatic or security engagement between India and Japan during the many decades of the Cold War, the deepening of Indo-Japanese ties during the past ten years can be

7. ... AND BEYOND TO THE PACIFIC

While looking East, India has also turned its gaze southward. Indo-Australian relations have recovered significantly from the diplomatic crisis perpetuated by India's 1998 nuclear tests. A series of annual bilateral talks begun in 2001 that focused attention on common security interests led to a renewed appreciation of the role that both countries could play in maintaining regional security. [131] As Indian strategist Raja Mohan has noted, Australia possesses untapped potential as an economic and strategic partner for India. [132] For their part, Australian leaders have recognized the important role that India can play in the security architecture of the wider Asia-Pacific region. [133] Australia strongly supported India's entry into the ASEAN Regional Forum and has publicly backed India's bid for a seat on the UN Security Council.

Bilateral agreements between the two nations have emphasized their "common interests on a number of important issues, including the Asia-Pacific and Indian Ocean regions." [134] This recognition of mutual interests led to a series of agreements in 2006 and 2007 on joint naval exercises, enhanced maritime security cooperation, increased military exchanges, and joint training of the two nations' armed forces. [135] Defense cooperation between the two countries also extends to research and development of military technology, as well as collaboration on counterterrorism efforts. Nuclear issues are an important aspect of Indo-Australian security ties as Australia has 40 percent of the world's uranium reserves. Nuclear cooperation received a significant boost when the government of John Howard decided to follow the Bush administration's lead in extending *de facto* recognition of India's nuclear status. [136] While not publicly endorsing its predecessor's policy, the Labor government of Kevin Rudd has been somewhat coy on this issue, leading some Australian analysts, such as Rory Medcalf, to believe that Australia will eventually supply uranium to India. [137] Despite this uncertainty, Indo-Australian security ties remain more robust than either nation's bilateral defense cooperation with China. With enhanced security ties to both Japan and Australia, India has assimilated itself with what Mohan calls the "northern anchor" and "southern anchor" of US military presence in Asia. [138]

India's increasing role in the Asia-Pacific has been firmly supported by the region's premier naval power – the United States. This has facilitated India's relations with the nations of the region because many Southeast Asian nations, as well as Japan, South Korea, and Australia, have close ties to America. Indo-American relations have evolved considerably from the days of the Cold War when India was viewed as a *de facto* Soviet client state. [139] Today, India and the United States share a range of concerns on key security issues facing the Asia-Pacific region such as the spread of Islamic radicalism, the proliferation of weapons of mass destruction, and the rise of China. [140] Furthermore, Indian leaders joined former President George W. Bush in advocating the spread of liberal democracy as a key element of long-term stability in Asia. [141]

From Washington's vantage point, India is poised to become a key player in world affairs. As a result, the Bush administration's policy had been to "help India become a major world power in the 21st century." [142] A cornerstone of this effort was the so-called "US–India nuclear deal" that would allow unprecedented civilian nuclear cooperation between the two nations. Although implementation of this agreement had been delayed by domestic politics within India, the direction and momentum of US–Indian defense cooperation is well established. [143] Between 2002 and

2006, the US and India conducted over 20 joint military exercises. [144] Moreover, a ten-year defense pact signed in June 2005 advanced intelligence sharing and the training of military personnel. [145] It also allowed military technology transfers, missile defense collaboration and arms sales, as well as opening the door to joint weapons production. When it comes to purchasing routine defense articles from the US, India presently receives the same status and treatment as long-time regional allies such as Japan and South Korea. [146] For example, the US has supported the sale of advanced defense materials to India, such as the Israeli-built *Phalcon* airborne early-warning aircraft, while opposing similar sales to China. [147] More recently, Lockheed Martin won a $1 billion contract to provide the Indian Air Force with Super Hercules C-130J military transport aircraft, and several American firms are bidding to supply the Indian military with advanced fourth-generation fighter jets and light helicopters. [148] Lockheed Martin has recently suggested that its cutting-edge 5th-generation strike aircraft, the F-35 Lightning II, could be made available to the Indian Air Force at some point in the future. [149]

With regard to China, there is a noteworthy similarity between Washington's and New Delhi's objectives vis-à-vis Beijing. Both nations have adopted "congagement" strategies that seek to gain from economic exchange with China while maintaining sufficient military power to deter threats to their key strategic interests posed by China's rising power. [150] This takes a tangible form in Southeast Asia, for example, where both India and the US seek to prevent the region from becoming economically and militarily dependent on China.

Former Secretary of State Condoleezza Rice favorably portrayed India as "a rising global power that can be a pillar of stability in a rapidly changing Asia," and the United States has encouraged New Delhi to take a greater role in the security of the Asia-Pacific region. [151] Toward that end, the US Pacific Command (PACOM) has sought to upgrade its regional coordination with India by inviting Delhi to post a liaison officer to PACOM headquarters in Hawaii – a privilege that had previously been confined to close regional allies such as Australia, South Korea, and Japan. [152] This development is not surprising when one reviews the 2006 Quadrennial Defense Review (QDR): in its high-level statement of US military strategy, the Department of Defense identifies India as a "key strategic partner" of the United States. As such, the QDR groups India in the same category as America's traditional Asia-Pacific allies. [153]

The American and Indian navies have conducted joint maneuvers in the Indian Ocean as part of the Malabar series of exercises since 1993. However, the first ever multilateral exercise featuring India, Japan, and the United States took place off of Tokyo Bay in mid-April 2007, featuring four Japanese guided missile destroyers, two American destroyers, and an Indian destroyer, corvette and tanker. A reciprocal exercise, Malabar 07-2, also involving Australia and Singapore, was held in the Bay of Bengal in September 2007. Featuring three aircraft carriers, 28 surface vessels, 150 aircraft and over 20,000 personnel, the five-day naval exercise was one of the largest ever held in the region. [154] Such multinational exercises increase both the Indian Navy's professional capabilities and its interoperability with friendly nations.

Despite these deepening ties, there do remain significant differences between India and the United States over a host of foreign policy issues ranging from Pakistan, and relations with Iran, to broader issues of global economic governance. [155] Moreover, at the time of this writing, it remains to be seen how the new Obama administration will approach the Indo-US relationship. Obama's initial opposition to the nuclear deal as a senator was a cause for concern, as was his presidential campaign's denunciation of then-rival Hillary Clinton as “the Senator from Punjab” during the Democratic Primary due to her numerous ties to India. These worries were compounded by the administration's subsequent failure to mention India on its list of foreign policy priorities, the decision to neglect New Delhi on the Secretary of State's inaugural trip to Asia (despite recommendations from the Policy Planning Staff that it be included), and the appointment of Richard Holbrooke, who has advocated an active US role in mediating the dispute over Kashmir, as special envoy – all of which raise questions as to whether the new administration was backing away from its predecessor's strategic engagement with India or was merely tone deaf in its decision-making. [156]

8. POTENTIAL CONSTRAINTS ON DELHI'S ASIAN VISION

Before moving on to an evaluation of India's impact on the regional order in Asia, it is necessary to balance the discussion in the previous sections by a look at the factors that could impede India's ability to develop as an extra-regional power. Despite its success to date, India still faces a number of challenges in its efforts to project its influence into the Asia-Pacific region. At the grand-strategic level, there are questions about India's ability to articulate and implement a coherent long-term national security strategy. Scholars both inside and outside India have found that its political establishment has difficulty approaching defense and foreign policy issues in a systematic manner. [157] Furthermore, there is not necessarily support for a robust Asia-Pacific role across the political spectrum. After vigorous protests by the left-wing parties over the multilateral nature of the 2007 Malabar exercise, the Indian government did not include Australia, Japan and Singapore in the 2008 version in an attempt to appease groups which had recently withdrawn their support from the country's governing coalition. [158] In the long run, these factors have the potential to hinder India's ability to integrate its political, military, and economic efforts to pursue its interests in the Asia-Pacific. [159]

In terms of military power, India is still at an early stage in developing its ability to project and sustain its military presence beyond the Indian Ocean. The development of new ports and airbases are a positive step in this direction. However, much of India's infrastructure remains underdeveloped and highly inefficient. In terms of military spending, India's defense budget ranks ninth in the world and is only the fifth largest in Asia behind the US, China, Japan, and South Korea. While the Indian Navy has been the recipient of record shares of the defense budget in recent years, receiving a 15 percent increase in procurement spending in 2008, achieving its ambitious acquisition goals will require sustaining present funding levels for many years – particularly given India's inefficient military industry. [160] Whether or not this can be achieved remains to be seen: not only does the Navy's budget face challenges from the Army and Air Force, which do not share the Navy's focus on maritime power-projection capabilities, but India's internal security situation could place demands on resources. Thus far the government has rightly refrained from employing the armed forces to counter the Naxilite communist insurgents, which Prime Minister Singh had labeled India's single biggest security threat; however, funding for internal defense could take priority should the situation worsen.

Although India's defense spending only accounts for 2.3 percent of GDP, the defense budget could face pressure from demands for increased social spending – particularly in light of the present global economic recession. [161] While India has recorded impressive economic growth over the past two decades, the country remains tremendously poor, with an estimated 22 percent of the population living below the international poverty line. [162] In particular, recent economic growth has largely bypassed the agricultural sector in which half of India's population is employed. In terms of GDP per capita, adjusted for purchasing power parity, India's $3,452 per person ranks 117th in the world. [163] In comparison to other Asian powers, the average Indian has a bit more than half the income of his Chinese counterpart and close to a tenth that of a Japanese citizen. Even among the emerging markets of Southeast Asia, India compares unfavorably to countries like Indonesia and the Philippines. In a democratic society, demands for investment in education, health care, and basic infrastructure may come at the expense of spending on defense and foreign affairs. Having long been defined by its poverty, India's standing as an Asian power depends, in part, on its internal development. [164]

A third challenge to Delhi's ability to focus its attention on the Asia-Pacific comes from what Stephen Cohen calls India's "Inner Ring." [165] India's immediate neighborhood contains several weak countries that run the risk of becoming failed states. [166] Furthermore, India's tense relationship with its nuclear-armed neighbor Pakistan has long been the central concern of Indian foreign and defense policy. Although successive Indian governments have taken active steps to move government attention away from single-minded focus on this sub-continental rivalry, Islamabad's continued support for terrorism within India and the very real threat of state failure in Pakistan necessarily draw India's attention westward. Similarly, the continued economic and political challenges facing the small, fragile states on India's periphery, such as Bangladesh, Sri Lanka, and Nepal, will require attention that could otherwise be given to developments in the Asia-Pacific. [167]

Though not insurmountable, the challenges posed by military capacity, economics, and immediate regional stability will necessarily constrain Delhi's attempts to enlarge India's role in the Asia-Pacific in the near term. Addressing this range of issues will require India's political establishment to marry effective leadership with the political will necessary to ensure that India's capabilities can eventually match its geopolitical ambition. Nevertheless, despite the handicaps, as the following section indicates, India is poised to influence the regional dynamics in the Asia-Pacific.

9. INDIA AND THE FUTURE OF THE ASIA-PACIFIC

An expanding economy, increasing maritime capability and security ties with Japan, Australia, and the United States, as well as key Southeast Asian nations, are positioning India to have an impact on the distribution of power in the Asia-Pacific. This is a role that successive Indian governments have actively cultivated – ambitiously expanding their strategic focus over the past two decades from South Asia to the Pacific. While India's foreign policy establishment may ultimately prefer to see the present unipolar system replaced with a multipolar one, in which India is a major power, a power transition in the Asia-Pacific region that leads to great power competition and increased instability is not in India's interests. Preferring *Pax Americana* to a Sino-centric world order, [168] India's Asia-Pacific policy has sought to enhance its own regional role while simultaneously seeking to hedge in its relations with its northern neighbor – benefiting from economic engagement while cultivating relationships with China-wary nations to match Beijing's perceived attempts at strategic encirclement in the Indian Ocean.

The emerging security structure of the Asia-Pacific region is far less certain than the relatively stable and transparent bipolar order that characterized the four-and-a-half decades of the Cold War. The salient characteristic of regional order in the past 15 years has been the preeminent position of the United States. American maritime power and its Cold War "hub and spoke" alliance structure have allowed it to exercise dominance while facilitating stability and economic growth throughout the region. However, the emergence of new powers in Asia such as India and China, and the increasing "normalization" of Japan as a political-military actor, appear ready to transform the Asia-Pacific. While America's military and political power in the region is hard to deny, its ideal regional order, based on the rule of law and democracy, is not necessarily attractive to all states. At the same time, China, which seeks to leverage its economic growth to either cast itself as an alternative to the US in a bipolar Asia or supplant America's regional role, has yet to make an attractive case for a Sino-centric order. [169] In such a dynamic environment, it is possible for other Asian powers to play an influential role in shaping regional security dynamics. The question of regional leadership in "the Asian Century" is not just a matter for the US and China alone, but for Japan, India, and the nations of Southeast Asia as well.

Although it is significantly beyond the scope of this article to definitively comment on the future security architecture of the Asia-Pacific, several potential regional orders can be identified which allow for a discussion of India's impact in shaping Asia's future. The first possible configuration is that of regional hegemony exercised by either the United States or China. A regional hegemon is not simply the preponderant regional power, as the US is in Asia today, but it is a state that is so powerful that "no other state has the military wherewithal to put up a serious fight against it." [170] China's expanding power and its possession of a nuclear arsenal render America's present regional position something short of hegemony. Conversely, America's presence in Asia prevents China from achieving hegemony. Even if the US were to significantly draw down its presence in the region, Japan and India together, which both oppose Chinese hegemony, possess enough combined power to prevent China from achieving a hegemonic position in the maritime Asia-Pacific. [171] While China may be widely acknowledged as the dominant land power in Asia, the region is fundamentally a maritime domain, and the PLAN lacks the ability to assert control over the strategic island chains in its own backyard, most notably Taiwan, let alone the rest of the Asia-Pacific region. [172] Furthermore, the ongoing dispute over Taiwan has led China to focus its naval modernization program on developing an access-denial capability in its territorial waters rather than on power projection. [173] The combination of Japan's naval forces, complemented by India's expanding expeditionary naval capability, presents sufficient naval power to deny Beijing true hegemony in the region. The balance tilts still further away from China if other regional navies are also taken into account.

With hegemony unlikely to emerge in Asia in the medium term, are either multipolar or bipolar structures likely to characterize the region's nascent security architecture? India would prefer a multipolar power structure in the Asia-Pacific. However, given the current preponderance of American power in the region, and the gap between the relative power of Japan, China, and India, it is unlikely that Asia will see the emergence of multiple poles of approximately equal power in the medium term. Furthermore, multipolarity suggests independence and balancing among the major states, which does not necessarily characterize US–Japan, US–India or Indo-Japanese ties. Similarly, the gap between American and Chinese power in Asia makes a balanced bipolar structure unlikely since significant actors such as Japan and India, as well as some less powerful Southeast Asian states, are more likely to support the United States (the stronger power) rather than China (the weaker one). [174]

As a result, the most likely configuration in Asia appears to be a continuation of the present: a hierarchical order with American preponderance. [175] Under such conditions, regional stability is preserved when the dominant power gains support for the status quo from other significant powers in the hierarchy that are satisfied with the present regional structure. This situation facilitates the maintenance of a power gap between the dominant state in the hierarchy and its supporters on the one hand and a would-be challenger on the other, thereby reducing the likelihood of great power conflict.

India recognizes the value of the existing US alliance system in providing stability in the Asia-Pacific region and shares the preferences of many states in East and Southeast Asia for maintaining US preponderance in the region via economic, political, and military engagement. [176] Although, at present, India lacks the ability to independently shape the regional order in the Asia-Pacific, it makes its presence felt by integrating itself with the major democracies of the region and expanding its ties with China-wary nations. In pursuing strategic ties with nations that have traditionally had difficult relations with China – such as Vietnam, Indonesia, Japan, and the United States – New Delhi lends its military and economic power to a regional security order that can enhance stability in Asia by presenting Beijing with a series of structural constraints that may diffuse the negative aspects of China's rise and persuade it that attempts to dominate the region are unlikely to succeed. In pursuing this course, India is not subordinating itself to another power or seeking to be a junior partner in any coalition; rather it is pursuing its own agenda as an emerging great power, whose interests presently coincide with those of the United States and its regional allies. Although it is possible that India's patterns of behavior and alignment described in this article could be reversed at some point in the future, the realities of geography, regional structure, and power dynamics in Asia make that unlikely.

10. CONCLUSION

Collectively, India's expanding economic ties, its growing partnerships with key regional actors, and its increasingly capable navy will have an important impact on the evolving regional order in the Asia-Pacific. After nearly half a century of "confinement" to the sub-continent, India is increasingly making its presence felt across Asia's various sub-regions, reprising a role that it played in centuries past. While this conclusion generally concurs with the thrust of recent assessments of India's foreign policy goals in Asia and its potential role in shaping the region's future security architecture by David Scott and Harsh Pant, it is somewhat more circumspect about India's present level of influence in the Asia-Pacific, which positions it closer to Scott's belief that "India has indeed become *a* Pacific player," than Pant's assertion that India is already "a *major* player in the Asia-Pacific regional balance of power." [177] In the near term, India's presence and influence will be felt most strongly in Southeast Asia, however it has clearly signaled an ambition to play a leading role in the international politics of the broader Asia-Pacific region.

The eastward focus, which has been a cornerstone of India's foreign policy since the end of the Cold War, is part of a broader effort to assert itself on the world scene. Over the past 18 years, India has evolved from a regional power in South Asia to an actor in the Asia-Pacific. Maintaining a significant gap between the power of the United States and its allies on the one hand and China on the other can help to deter Beijing from mounting a costly bid for regional hegemony, which, successful or not, would increase instability throughout the Asia-Pacific. It is in India's interests, as well as that of many states in East and Southeast Asia, to avert a power transition in the region. In so far as India continues to contribute to that effort through its increasingly capable Navy, strategic partnerships with key regional actors, and growing trade and investment links, it will play an important role in shaping regional dynamics in the Asia-Pacific. Although it will be some time, if ever, before India's power projection and political influence match the full extent of its regional ambition, it is clear that India is much more than a state that merely "interacts with the Asia-Pacific in various ways." [178]

For their helpful comments on earlier drafts of this article, the author would like to thank Andrew Erickson, Matthew Jenkinson, Catherine Lang, Manjeet Pardesi, David Scott, Daniel Twining, and the three anonymous reviewers for *Asian Security*.

NOTES

1. Derek McDougall, *Asia-Pacific in World Politics* (Boulder, CO: Lynne Rienner Publishers, 2007), p. 7.
2. Harsh Pant, "India in the Asia-Pacific: Rising Ambitions with an Eye on China," *Asia-Pacific Review* Vol. 14, No. 1 (2007), p. 57.
3. Satu P. Limaye, "India–East Asia and U.S.–India Relations: Movin' on Up," *Comparative Connections* Vol. 8, No. 4 (January 2007), p. 166.
4. Chinese Ministry of Commerce, "Top Ten Trading Partners," March 21, 2008. Available at http://english. mofcom.gov.cn/aarticle/statistic/ie/200803/20080305439972.html.
5. "Full text of joint statement of China, India," *People's Daily*, April 13, 2005.

6. Stephen J. Blank. *Natural Allies? Regional Security in Asia and Prospects for Indo-American Strategic Cooperation* (Carlisle, PA: Strategic Studies Institute, 2005), p. 66.

7. "Memorandum of Understanding Between the Ministry of Defense of the Republic of India and the Ministry of National Defense of the People's Republic of China for Exchanges and Cooperation in the Field of Defense," May 29, 2006.

8. Manjeet S. Pardesi, "The India–China Border," in Clinton Fernandes, ed., *Hot Spot: Asia and Oceana* (London: Greenwood, 2008), pp. 67–89.

9. See "China, India fight for African Oil," *Agence France-Presse*, October 15, 2004 and "India Loses PetroKaz Bid to China," *The Times of India*, August 23, 2005.

10. Šumit Ganguly, "Energy Trends in China and India," July 26, 2005, p. 3. Available at http://lugar.senate. gov/energy/hearings/pdf/050726/GangulyTestimony.pdf.

11. Sudha Ramachandran, "China Moves into India's Back Yard," *Asia Times*, March 13, 2007.

12. Scholars of Chinese maritime issues are not convinced that the "string of pearls" concept is an accurate conception of Beijing's strategy – in particular they note that the phrase rarely appears in Chinese naval writings. Author's correspondence with Dr. Andrew S. Erickson, China Maritime Studies Institute, US Naval War College, October 9, 2007. For an assessment that the PLAN may attempt to gain access to Chinese built dual-use port facilities in the Indian Ocean, but it unlikely to develop full-scale bases in the medium term, see Gurpreet S. Khurana, "China's 'String of Pearls' in the Indian Ocean and Its Security Implications," *Strategic Analysis* Vol. 32, No. 1 (January 2008), pp. 1–39.

13. Tarique Niazi, "Gwadar: China's Naval Outpost on the Indian Ocean," *China Brief* Vol. 5, No. 4 (February 14, 2005); and Sudha Ramachandran, "China's Pearl in Pakistan's Waters," *Asia Times*, March 4, 2005.

14. Gurmeet Kanwal, "Countering China's Strategic Encirclement of India," *Indian Defence Review* Vol. 15, No. 3 (July–September 2000), p. 13; and C. S. Kuppuswamy, "Myanmar–China

Cooperation: Its Implications for India," *South Asia Analysis Group*, February 3, 2003. Available at http://www.saag. org/papers6/ paper596.html. For a skeptical assessment, see Andrew Selth, "Chinese Military Bases in Burma: The Explosion of a Myth," *Regional Outlook Paper No. 10* (Nathan, Australia: Griffith Asia Institute, 2007).

15. Ramachandran, "China Moves into India's Back Yard."

16. Brahma Chellaney, "Forestalling Strategic Conflict in Asia," *Far Eastern Economic Review* Vol. 169, No. 9 (November 2006), pp. 29–33; and John W. Garver, *Protracted Contest: Sino-Indian Rivalry in the Twentieth Century* (Seattle, WA: University of Washington Press, 2001), p. 5. The Indian Navy's maritime doctrine explicitly discusses "attempts by China to strategically encircle India" and warns of Chinese encroachment into "our maritime zone." Cited in "India's Naval Posture: Looking East," *Strategic Comments*, Vol. 11, No. 6 (August 2005), p. 2.

17. Vijay Sakhuja, "Indian Navy: Keeping Pace with Emerging Challenges," in Lawrence W. Prabhakar et al., eds., *The Evolving Maritime Balance of Power in the Asia-Pacific* (Singapore: World Scientific, 2006), p. 191 and "Power Realignments in Asia: China, India and the United States" (New Delhi: Observer Research Foundation, December 14–17, 2006), p. 8. The PLAN's near-term ability to project and sustain power into the Indian Ocean has been questioned by some scholars. James Holmes and Toshi Yoshihara, "China's Naval Ambitions in the Indian Ocean," *Journal of Strategic Studies* Vol. 31, No. 3 (June 2008), pp. 388–391.

18. "Finally, Pranab Calls China a Challenge," *Times of India*, November 5, 2008.

19. Baldev R. Nayar and T.V. Paul, *India in the World Order: Searching for Major Power Status* (Cambridge: Cambridge University Press, 2003), p. 230.

20. Ashok Kapur, *India – From Regional to World Power* (London: Routledge, 2006), pp. 219–223.

21. Discussion of the views of Indian policymakers towards China is drawn from Mohan Malik, "Eyeing the Dragon" (Honolulu, HI: Asia-Pacific Center for Security Studies, December 2003.)

22. Pew Research Center, "Global Economic Gloom – China and India Notable Exceptions," *Pew Global Attitudes Project*, June 12, 2008, p. 43.

23. Pew Research Center, "U.S. Image Up Slightly, But Still Negative: American Character Gets Mixed Reviews," *Pew Global Attitudes Project*, June 23, 2005, p. 33.

24. Garver, *Protracted Contest*; Mohan Malik, "India–China Relations" (Honolulu: Asia-Pacific Center for Security Studies, October 2004), p. 18; Šumit Ganguly, "India and China: Border Issues, Domestic Integration, and International Security," in Francine R. Frankel and Harry Harding, eds., *The India–China Relationship: What the United States Needs to Know* (New York: Columbia University Press, 2004), pp. 103–133; and Ashley J. Tellis, "China and India in Asia," in Frankel and Harding, eds., *The India–China Relationship*, pp. 134–177.

25. Chung Chien Peng, *Domestic Politics, International Bargaining, and China's Territorial Disputes* (London: Routledge, 2004), p. 17. See also Guihong Zhang, "US–India Strategic Partnership: Implications for China," *International Studies* Vol. 42, No. 3–4 (October 2005), p. 289.

26. Ashley Jackson, "The British Empire in the Indian Ocean," in Dennis Rumley and Sanjay Chaturvedi, eds., *Geopolitical Orientations, Security and Regionalism in the Indian Ocean* (New Delhi: South Asian, 2004), p. 35.

27. James Morris, *Farewell the Trumpets: An Imperial Retreat* (San Diego: Harcourt Brace Jovanovich, 1978), p. 129–130.

28. George N. Curzon, *The Place of India in the Empire* (London: J. Murray, 1909), p. 12.

29. For example, veteran of the Indian Civil Service and former Foreign Secretary of the Raj, Sir Olaf Caroe firmly believed that, "in the modern world it is inevitable for India to be the centre of affairs of Asia." A. G. Noorani, "Caroe's Lessons," *Frontline* Vol. 23, No. 9 (May 6, 2006).

30. This policy has been characterized as neo-Curzonian – after the British imperial viceroy. C. Raja Mohan, *Crossing the Rubicon* (New York: Palgrave MacMillan, 2003).

31. "India challenges China in South China Sea," *Asia Times*, April 27, 2000. A similar view of India's "extended neighborhood" was articulated by the National Security Advisor Brajesh Mishra in "Global Security: An Indian Perspective," *Speech presented at the National Defence Institute*, Lisbon, April 13, 2000. Available at http://meaindia.nic.in/disarmament/dm13apr00.htm.

32. Integrated Headquarters (Navy), *Indian Maritime Doctrine* (New Delhi: Ministry of Defense, April 2004), p. 56.

33. Mohan explicitly cites Curzon in arguing that India's naval focus should go beyond the Indian Ocean littoral. Mohan, *Crossing the Rubicon*, pp. 204–7.

34. Manjeet S. Pardesi, "Deducing India's Grand Strategy of Regional Hegemony from Historical and Conceptual Perspectives" (Singapore: IDSS, 2005), p. 55.

35. Manmohan Singh, "Prime Minister's Address," Speech presented at the Combined Commanders Conference, New Delhi October 26, 2004. Available at http://pmindia.nic.in/lspeech.asp?id=37. See also, Shyam Saran, "Present Dimensions of the Indian Foreign Policy," Speech presented at the Shanghai Institute of International Studies, Shanghai, January 11, 2006. Available at http://www.indianembassy.org.cn/press/fsshanghaispeech.htm

36. Arun Prakash, "Shaping India's Maritime Strategy," Speech presented at Indian National Defense College, November 2005. Available at http://indiannavy.nic.in/cns_add2.htm.

37. "Indian Navy Modernization: 6 New Submarines, 33 Ships to be Acquired," *India Defence*, July 1, 2007. Available at http://www.india-defence.com/reports/3384.

38. "Indian Navy Chief Admiral Sureesh Mehta Spells Out Vision 2022," *India Defence*, August 10, 2008. Available at http://www.india-defence.com/reports/3954.

39. James Hackett, ed., *The Military Balance, 2008* (London: International Institute for Strategic Studies, 2008), p. 336.

40. Hackett, *The Military Balance, 2009*, p. 335.

41. Kailash Kohli, "Naval Gazing into the Future," *The Indian Express*, December 13, 2003.

42. In contrast, for an external assessment that concurs with the Indian Navy's more optimistic delivery timelines, see Andrew C. Winner, "India as a Maritime Power?," in Toshi Yoshihara and James Holmes, eds., *Asia Looks Seaward: Power and Maritime Strategy* (Westport, CT: Praeger, 2008), p. 141.

43. Rajat Pandit, "After Gorshkov, Another Navy Project Hit by Delay," *Times of India*, August 10, 2007.

44. Rahul Bedi, "Indian Navy Strives for Regional Dominance," *Jane's Defense Weekly*, December 21, 2005.

45. Arunkumar Bhatt, "Focus on Navy's Influence on Land Operations," *The Hindu*, May 7, 2006.

46. Barry R. Posen, "Command of the Commons: The Military Foundation of U.S. Hegemony," *International Security* Vol. 28, No. 1 (Summer 2003), p. 11.

47. Rajat Pandit, "Defending India: An Ageing Naval Fleet, Obsolete Equipment," *Times of India*, October 25, 2008.

48. "Russian Sub in Accident was Intended for India," *Associated Press*, November 10, 2008.

49. Hackett, *Military Balance, 2009*, p. 339.

50. Rajat Pandit, "Navy Hunts for Hi-Tech Submarines," *Times of India*, October 20, 2008.

51. Rajat Pandit, "India Inks Largest-Ever Defence Deal with U.S.," *Times of India*, January 5, 2009.

52. Capable of achieving Mach 2.5, the BraMos is three times faster than the US Tomahawk cruise missile.

53. Sureesh Mehta, "India's Maritime Diplomacy and International Security," Speech presented at the International Institute for Strategic Studies, June 21, 2007.

54. "Russia May Build More Krivak Class Frigates for India," *Novosti* (Russia), February 4, 2009.

55. "Indian Navy Seeks Second Troop Carrier from US," *News Post India*, July 14, 2007. Available at http://www.newspostindia.com/report-7326.

56. Vijay Sakhuja, "Emerging Contours of Asian Naval Power," *Opinion Asia*, February 5, 2007.

57. Some observers have suggested that the simultaneous buildup of naval power by China, India, Japan, South Korea and others is evidence of an arms race in the Asia-Pacific. "Into the wide blue yonder: Asia's main powers are building up their navies. Is this the start of an arms race?," *The Economist*, June 7, 2008, p. 53. While it is beyond the scope of this article to assess the casual patterns of weapons acquisition by those nations, it is worth noting that, from a historical standpoint, emerging great powers tend to build navies.

58. C. S. Kuppuswamy, "India's Look East Policy: More Aggressive, Better Dividends," *South Asia Analysis Group*, January 3, 2006. Available at http://www.saag.org/%5Cpapers17%15Cpaper1663.html.

59. G. V. C. Naidu, "Whither the Look East Policy: India and Southeast Asia," *Strategic Analysis* Vol. 28, No. 2 (April–June 2004), p. 338.

60. "Instrument of Accession to the Treaty of Amity and Cooperation in Southeast Asia," Bali, October 8, 2003. Available at http://www.aseansec.org/15282.htm.

61. Zhao Hong, "India and China: Rivals or Partners in Southeast Asia?," *Contemporary Southeast Asia* Vol. 29, No. 1 (April 2007), p. 139.

62. "India to Boost Trade Ties with SE Asia under New 'Look East' Focus," *Channel News Asia* (Singapore), November 11, 2005; Amitav Acharya, "Will Asia's Past Be Its Future?," *International Security* Vol. 28, No. 3 (2003/04), pp. 150–151; "Thailand Looks to India to Balance China's Economic Weight," *Agence France-Presse*, June 24, 2007.

63. Amit Baruah, "India has Legitimate Interests in South East Asia: George Yeo," *The Hindu*, January 24, 2007.

64. Lee Hsien Loong, "The Future of Asian Economies," Speech presented at the Davos World Economic Forum, January 24, 2003.

65. Kuppuswamy, "India's Look East Policy."

66. Acharya, "Will Asia's Past Be Its Future?," pp. 150–151.

67. Quoted in Simon Elegant and Michael Elliot, “Lee Kuan Yew Reflects,” *Time Asia*, December 5, 2005; Mahendra Ved, “Malaysia–India Ties Set for Next Level,” *New Straits Times* (Malaysia), February 18, 2007.

68. K. Kesavapany, “Dire Need for an Asian Renaissance,” *Business Times* (Singapore), March 28, 2008; Meidyatama Suryodiningrat, “Is India Ready to be Part of Southeast Asia Again?” *Jakarta Post*, June 18, 2007.

69. Vishal Thapar, “Navy Deploys Warships in South China Sea,” *Hindustan Times*, October 18, 2004; P. S. Suryanaryana, “Aircraft Carrier on a Friendly Mission,” *The Hindu*, August 3, 2005. Not only does the Indian Navy view its aircraft carrier as an important symbol of its maritime power, it contrasts favorably with China which lacks a similar capability. For a discussion of the latter point, see Andrew S. Erickson and Andrew R. Wilson, “China's Aircraft Carrier Dilemma,” *Naval War College Review* Vol. 59, No. 4 (Autumn 2006), pp. 13–45.

70. “Indian Navy Engages U.S. and Russia Away From Home,” Press Release, Press Information Bureau, Government of India, March 29, 2007. Available at http://pib.nic.in/release/release.asp?relid=26495.

71. Stanley Chan, “Quiet Power: ASEAN,” *Asia Times*, August 10, 2002.

72. Ramachandran, “IndiaNavy Drops Another Anchor.”

73. “Wary of China, India to Boost Eastern Naval Fleet,” *The Indian Express*, November 15, 2007.

74. Maitra, “India Bids to Rule the Waves.”

75. K. Dhar, “Indian Air Force Carries out Exercises from Andaman Islands Base,” *Press Trust of India*, April 15, 2005.

76. Sudha Ramachandran, “India Signs on as Southeast Asia Watchdog,” *Asia Times*, April 5, 2002.

77. C. Raja Mohan, “Looking East: Phase Two,” *The Hindu*, April 11, 2002.

78. Marwaan Macan-Markar, "India Shifts Regional Geopolitical Cards," *Asia Times*, January 27, 2005; Rajat Pandit, "Navy Makes a 'Blue-Water' Mark," *The Times of India*, January 7, 2005.

79. "India Ready to Play Role in Protecting Malacca Straits," *Press Trust of India*, June 17, 2007.

80. Suryanaryana, "Aircraft Carrier on a Friendly Mission."

81. Tim Huxley, *Defending the Lion City: The Armed Forces of Singapore* (St. Leonards: Allen and Unwin, 2000), pp. 33–34.

82. Kuppuswamy, "India's Look East Policy"; "Vietnam–India Strategic Partnership Anniversary Celebrated," *Vietnam News Agency*, July 5, 2008.

83. Some hawkish Indian analysts see close ties with Vietnam as payback for China's support of Pakistan. "Vietnam Keen on Forging New Strategic Equations," *Hindustan Times*, June 27, 2007.

84. "Viet Nam, India Issue Joint Declaration on Strategic Partnership," New Delhi, India, July 6, 2007. Available at http://www.mofa.gov.vn/en/nr040807104143/nr040807105001/ns07070909122.

85. Between China, the US and India, Storey and Thayer conclude that India is most likely to be granted access privileges. Ian Storey and Carlyle Thayer, "Cam Ranh Bay: Past Imperfect, Future Conditional," *Contemporary Southeast Asia* Vol. 23, No. 3 (December 2001), pp. 452–473.

86. "There is Political, Economic Space for all Countries," *Jakarta Post*, June 15, 2007; "Mapping the Global Future" (Washington, DC: National Intelligence Council, 2004), p. 29. In terms of GDP per capita, even adjusted for purchasing power parity, Indonesia will still lag far behind major developed economies.

87. Kuppuswamy, "India's Look East Policy."

88. "India, Indonesia Mull Joint Defense Hardware Production," *India Defence*, June 15, 2007. Available at http://www.india-defence.com/reports/3311.

89. Abdul Khalik, "Indonesia–India Security Pact Comes into Effect," *The Jakarta Post*, April 3, 2007.

90. Sushil Seth, "China Factor in India-Indonesia Ties," *Taipei Times*, April 20, 2007.

91. "India, Malaysia Hold Joint Naval Exercises," *Times of India*, May 3, 2008.

92. "India Ready to Sell Supersonic Bramos Missile to Malaysia," *Malaysia Economic News*, October 4, 2007.

93. "Malaysia's Foreign Minister Calls for 'Strategic Alliance,'" *Bernama* (Malaysia), February 14, 2007.

94. "Thais, Indians in Anti-Terror Exercise," *Bangkok Post*, September 10, 2007.

95. "Philippines to Sign Defence Pact with India," *India Defence* February 1, 2006. Available at http://www. india-defence.com/reports/1284.

96. Arthur Waldron, "The Rise of China: Military and Political Implications," *Review of International Studies* Vol. 31, No. 4 (October 2005), p. 723.

97. Acharya, "Will Asia's Past Be Its Future?," p. 150; Hong, "India and China," p. 139.

98. "India Challenges China in South China Sea," *Stratfor.com*, April 27, 2000.

99. C. Raja Mohan and Parag Khanna, "Getting India Right," *Policy Review*, No. 135 (February–March 2006) . Available at http://www.hoover.org/publications/policyreview/2913806.html.

100. Horimoto Takenori, "The World as India Sees It," *Gaiko Forum: Japanese Perspectives on Foreign Affairs* Vol. 6, No. 3 (Fall 2006), p. 6.

101. V. Jayanth, "India and ASEAN Set to Focus on East Asia," *The Hindu*, January 12, 2007. For a broader discussion of the hedging strategies of Southeast Asian states vis-à-vis China, see Yuen Foong Khong, "Coping with Strategic Uncertainty: The Role of Institutions and Soft Balancing in Southeast Asia's Post-Cold War Strategy," in J. J. Suh et al., eds., *Rethinking Security in East Asia: Identity, Power and Efficiency* (Stanford, CA: Stanford University Press, 2004), pp. 172–208; and Amitav Acharya and See Seng Tan, "Betwixt Balance

and Community: America, ASEAN, and the Security of Southeast Asia," *International Relations of the Asia Pacific* Vol. 6, No. 1 (2006), pp. 37–59.

102. Yashwant Sinha, "Resurgent India in Asia," Speech presented at Harvard University, 29 September 2003.

103. E. Ahamed, "Reinforcing 'Look East' Policy," January 17, 2006. Available at http://mea.gov.in/interview/2006/01/17in01.htm.

104. Mohan Malik, "China and the East Asian Summit: More Discord than Accord" (Honolulu, HI: Asia-Pacific Center for Security Studies, February 2006).

105. "East Asia Summit: In the Shadow of Sharp Divisions," *People's Daily*, December 5, 2005.

106. Kuppuswamy, "India's Look East Policy." A similar point is made in Malik, "China and the East Asian Summit," p. 3.

107. C. Raja Mohan, "Look East Policy: Phase Two," *The Hindu* October 9, 2003.

108. "India, South Korea Military Exercise in July," *India Defence*, March 27, 2006. Available at http://www. india-defence.com/reports/1624.

109. Rajeev Sharma, "India, South Korea Join Hands Against Terrorism," *The Tribune*, October 7, 2004.

110. "Joint Exercises with S. Korea in March 2006," *India Defence*, September 27, 2005. Available at http://www. india-defence.com/reports/436.

111. Mohan Malik, "The Proliferation Axis: Beijing–Islambad–Pyongyang," *The Korean Journal of Defense Analysis* Vol. 15, No. 1 (Spring 2003), pp. 57–98.

112. Malik, "The Proliferation Axis," p. 59.

113. Rajeev Srinivasan, "An Indo-Japanese Strategic Alliance," *Rediff.com*, 26 April 2005. Available at http://in.rediff.com/news/2005/apr/2026rajeev.htm. The term "alliance" is sensitive in Indian politics, and is not an accurate description of this emerging partnership.

114. Joseph S. Nye, "The Rise of Liberal Japan," *Chosun Ilbo* (Seoul), June 15, 2007.

115. Richard J. Samuels, "New Fighting Power: Japan's Growing Maritime Capabilities and East Asian Security," *International Security* Vol. 32, No. 3 (Winter 2007/08), p. 111.

116. Harold Brown et al. "Chinese Military Power" (New York: Council on Foreign Relations, 2003), p. 2; and MichaelRichardson, "China Military Build-up Scares Asian Neighbors," *Jakarta Post*, April 17, 2007. Similar concerns are expressed in Japan's 2005 defense white paper, *Defense of Japan 2005* (Tokyo: Japan Defense Agency, 2005), p. 14.

117. "Bridging the India-China Gap," *Japan Times*, November 1, 2008.

118. On the importance of the differentiation between democracies and autocracies in contemporary international politics, see Robert Kagan, "End of Dreams, Return of History," *Policy Review*, No. 144 (August–September 2007).

119. "Japan–India Partnership Vital in East Asia," *Yomiuri Shimbun* (Tokyo), December 15, 2006.

120. Interestingly, it has been reported that Delhi was initially "lukewarm about the quad until Beijing began lobbying against it, leading India to sign on enthusiastically." Daniel Twining, "Playing the America Card," *The Weekly Standard* Vol. 13, No. 3 (October 1, 2007). Available at http://www.weeklystandard.com/ Content/ Public/Articles/000/000/014/146xmjrl.asp.

121. Prasan K. Gupta, "Looking East: India Forges Closer Ties with Japan," *Force* Vol. 3, No. 3 (November 2005). Available at http://www.forceindia.net.

122. Lalit Mansingh, "India–Japan Relations," *IPCS Issue Brief*, No. 43 (January 2007), p. 2.

123. "Joint Declaration on Security Cooperation between Japan and India," Tokyo, October 22, 2008.

124. Siddharth Srivastava, "China Looks on at the US–India Lockstep," *Asia Times*, June 30, 2007.

125. S. D. Naik, "India–Japan Ties – Moving to the Next Level," *The Hindu Business Line*, January 2, 2007.

126. Mansingh, "India–Japan Relations."

127. "Japan Courts India to Counter China," *Agence France-Presse*, August 23, 2007. Similarly, Reuters reports that Indo-Japanese linkages are being cultivated specifically to counter China's diplomatic and economic influence in Asia. George Nishiyama, "Japan PM Visits India, Eying Trade and China," *Reuters*, August 21, 2007.

128. Brama Chellaney, "Japan–India Partnership Key to Bolstering Stability in Asia," *The Japan Times*, December 14, 2006. The dynamic nature of the naval balance in Asia is taken up in a provocative manner by Robert D. Kaplan, "Lost at Sea," *New York Times*, September 21, 2007.

129. Amit Mukherjee, "From Japan with Love," *Business Today*, September 23, 2007.

130. Subhash Kapila, "Japan–India Joint Declaration on Security Cooperation (2008) Overhyped," *South Asia Analysis Group*, November 13, 2008. Available at http://www.southasiaanalysis.org/%5Cpapers30%5Cpaper 2921.html.

131. Anthony Bergin, "Benefits for Both Sides in Close Ties with India," *The Australian*, August 22, 2001.

132. Mohan, "Look East Policy."

133. Stephen Smith "Australia and India: A New Partnership in the Asia Pacific Century," Speech presented at Indian Council of World Affairs, September 11, 2008.

134. "India–Australia Strategic Dialogue," Australian High Commission, New Delhi, August 30, 2001. Available at http://www.india.embassy.gov.au/ndli/PA_12_01.html.

135. "Australia Inks First Defence Pact with India," *India Defence*, March 7, 2006. Available at http://www.india-defence.com/reports/1452; Mark Dodd, "India Defence Ties to be Tightened," *The Australian*, June 4, 2007.

136. Kaushik Kapisthalam, "Australia and Asia's Rise," *Australian Journal of International Affairs* Vol. 60, No. 3 (September 2006), p. 371.

137. Rory Medcalf, "Don't Call it Blackmail," *Indian Express*, September 10, 2008.

138. C. Raja Mohan, "The Asian Balance of Power," *Seminar*, No. 487 (March 2000). Available at http://www. india-seminar. com/2000/2487/2487%2020raja%2020mohan.htm.

139. Anit Mukherjee, "Curzon's Ghost: The Making of India's New Foreign Policy," *Armed Forces Journal*, May 3, 2006. Available at http://www.armedforcesjournal.com/story.php?F=1657826_0501.

140. Mohan and Khanna, "Getting India Right"; Sinha, "Resurgent India in Asia."

141. "India–U.S. Global Democracy Initiative," Indian Embassy, Washington, DC, July 18, 2005. Available at http://www.indianembassy.org/press_release/2005/July/15.htm

142. "Background Briefing by Administration Officials on U.S.–South Asia Relations," US Department of State, March 25, 2005. Available at http://www.state.gov/r/pa/prs/ps/2005/43853.htm. For a discussion of this policy and its implications, see Daniel Twining, "America's Grand Design in Asia," *The Washington Quarterly* Vol. 30, No. 3 (Summer 2007), pp. 79–94.

143. Author's interview with US Defense Department official, September 19, 2007. This optimism in the long-term trends in Indo–US relations is shared by Indian policymakers. Peter R. Lavoy and Robin Walker, "U.S. India Strategic Partnership: A Track-Two Dialogue for Long-Term Cooperation" (New Delhi: Institute for Defense Studies and Analysis, April 26, 2007), pp. 3–4.

144. Anupam Srivastava, "India: Toward True Partnership," *The Journal of International Security Affairs* No. 11 (Fall 2006), p. 26.

145. For example, the US Navy trained Indian fighter pilots to operate from aircraft carriers so that they can man the *INS Vikramaditya* when it comes into service. "Navy Pilots Head for 'Top Gun' Training in the US," *Times of India*, November 1, 2005.

146. Robert D. Blackwill, "US–India Defense Cooperation," *The Hindu*, May 13, 2003.

147. "India Signs AWACS Deal Worth $1bn," *The Times of India*, October 11, 2003.

148. "Lockheed wins $1bn Indian order," BBC News, February 18, 2008. Available at http://news.bbc.co.uk/2/hi/business/7250474.stm; "Fact Sheet: U.S.–India Defense Relationship" (Washington, DC: US Department of Defense, March 2006); and "Boeing Set to Bag Order for 8 Patrol Aircraft," *The Financial Express* (Mumbai), March 17, 2008.

149. "Report: India could get Joint Strike Fighter from US," *The International Herald Tribune*, July 19, 2007.

150. For the origins of the phrase "congagement" see Zalmay Khalilzad, "Congage China," *RAND Issue Paper IP-187* (Santa Monica, CA: Rand, 1999).

151. Condoleezza Rice, "U.S.–India Civilian Nuclear Cooperation Agreement," April 5, 2006. Available at http://www. state.gov/secretary/rm/2006/64146.htm. India's activities in the Western Pacific are discussed in David Scott, "Strategic Imperatives of Indiaas an Emerging Player in Pacific Asia," *International Studies* Vol. 44, No. 2 (2007), pp. 133–134.

152. "US plans Indian Navy Liaison officer in Pacific Command," *The Financial Express* (Mumbai), April 16, 2005.

153. *Quadrennial Defense Review Report* (Washington, DC: Office of the Secretary of Defense, 2006), p. 28.

154. "Malabar 2007: India, United States, Japan, Australia, Singapore Begin Massive 5-Day Naval Exercises," *India Defence*, September 3, 2007. Available at http://www.india-defence.com/reports/3519.

155. Šumit Ganguly, "India in 2008: Domestic Turmoil and External Hopes," *Asian Survey* Vol. 49, No. 1 (January/February 2009), p. 48.

156. The Agenda: Foreign Policy. Available at http://www. whitehouse.gov/agenda/foreign_policy. The State Department Policy Planning Staff had recommended that India, along with Japan, China, South Korea and Indonesia, make up the itinerary for the Secretary of State's first trip abroad, but India was subsequently not

included on the list. Author's interview with former US State Department official, March 1, 2009.

157. Jaswant Singh, *Defending India* (New York: St.Martin's Press, 1999), pp. 1–58; George Tanham, *Indian Strategic Thought* (Santa Monica, CA: Rand, 1992).

158. "Antony Nixes Expansion of India-US Malabar Naval Exercise," *Indo-Asian News Service*, September 24, 2008.

159. Baladas Ghoshal, "Some New Thoughts on India's Look East Policy," *IPCS Issue Brief* No. 54 (October 2007), p. 1.

160. Hackett, *Military Balance, 2009*, p. 339; Timothy Hoyt, *Military Industry and Regional Defense Policy* (Oxford: Routledge, 2007), pp. 22–66.

161. Hackett, *Military Balance, 2009*, p. 449.

162. Šumit Ganguly and Manjeet S. Pardesi, "India Rising: What is New Delhi to Do?," *World Policy Journal* Vol. 24, No. 1 (Spring 2007), p. 10.

163. *United Nations Human Development Report 2007/2008* (New York: Palgrave Macmillan, 2007).

164. For a discussion of additional social factors impacting India's development, see Ganguly and Pardesi, "India Rising," pp. 13–14.

165. Stephen Philip Cohen, *India: Emerging Power* (Washington, DC: Brookings Institution Press, 2001), p. 232.

166. *Foreign Policy* magazine's 2007 index of failed states included five of India's neighbors in the top 25: Afghanistan, Pakistan, Bangladesh, Nepal and Sri Lanka. "The Failed States Index 2007," *Foreign Policy*, No. 161 (July/August 2007), pp. 54–63.

167. Cohen, *India*, p. 266.

168. Evelyn Goh, *Meeting the China Challenge* (Washington, DC: East–West Center, 2005), p. 1.

169. Liselotte Odegaard, "The Balance of Power in Asia-Security: U.S.–China Policies on Regional Order," *The Korean Journal of Defence Analysis* Vol. XIX, No. 1 (Spring 2007), pp. 30–36. For arguments that China will seek to displace the US in Asia, see John

J. Mearshimer, *The Tragedy of Great Power Politics* (New York: W. W. Norton, 2001), pp. 360–402; and Aaron L. Friedberg, "Ripe for Rivalry: Prospects for Peace in a Multipolar Asia," *International Security* Vol. 18, No. 3 (Winter 1993/1994), pp. 5–33.

170. Mearsheimer, *Tragedy*, p. 40.

171. The spiraling cost of weapons procurement may drive down the size of the US Navy in the future, which will impact its presence in the Asia-Pacific. *Report to Congress on Annual Long-range Plan for Construction of Naval Vessels for FY 2009* (Washington, DC: Office of the Chief of Naval Operations, February 2008), pp. 3–12.

172. For an assessment by the PLA that Taiwan constrains China's maritime power projection, see Peng Guangqian and Yao Youzhi, eds., *The Science of Military Strategy* (Beijing: Military Science Publishing House, 2005), p. 443.

173. Andrew S. Erickson, "Can China Become a Maritime Power?," in Yoshihara and Holmes, eds., *Asia Looks Seaward*, pp. 70–110.

174. Goh, *China Challenge*, p. 1.

175. For the original treatment of hierarchies in international politics, see A. F. K. Organski, *World Politics* (New York: Alfred A. Knopf, 1958).

176. Michael Richardson, "Naval Contemplation," *South China Morning Post*, December 29, 2008.

177. Scott, "Strategic Imperatives," p. 136; Pant, "India in the Asia-Pacific," p. 57. (Emphases added.)

178. McDougall, *Asia-Pacific in World Politics*, p. 7.

CHAPTER 9

Generating test cases for marine safety and security scenarios: a composition framework

Hamed Yaghoubi Shahir1*, Uwe Glässer1, Roozbeh Farahbod2*, Piper Jackson1 and Hans Wehn3

1Software Technology Lab, Computing Science, Simon Fraser University, B.C., Canada

2SAP Research, Karlsruhe, Germany

3MDA Corporation, Research & Development, Richmond, B.C., Canada

ABSTRACT

In this paper we address the problem of testing complex computer models for infrastructure protection and emergency response based on detailed and realistic application scenarios using advanced computational methods and tools. Specifically, we focus here on testing situation analysis decision support models for marine safety & security operations as a sample application domain. Arguably, methodical approaches for analyzing and validating situation analysis methods, decision support models, and information fusion algorithms require realistic *vignettes* that describe in great detail how a situation unfolds over time depending on initial configurations, dynamic environmental conditions and uncertain operational aspects. Meaningful results from simulation runs require appropriate test cases, the production of which is in itself a complex activity. To simplify this task, we introduce here the conceptual design of a Vignette Generator that has been developed and tested in an industrial research project. We also propose a framework for composing vignettes from reusable vignette elements together with a formal representation for vignettes using the Abstract State Machine method and illustrate the approach by means of various practical examples.

KEYWORDS

Marine Safety & Security Test Case Generation Infrastructure Protection Emergency Response Abstract State Machines

1 INTRODUCTION

Infrastructure protection and emergency response scenarios routinely call for intelligent coordination and management of multiple mobile resources, often operating in vast geographical environments. For instance, mobile platforms including patrol and cargo airplanes, rescue helicopters, unmanned aerial vehicles (UAV), satellites, ground vehicles, and search & rescue (SAR) vessels are deployed for the gathering of information that is crucial for situation analysis and for the transportation of persons and supplies. Typical scenarios include: intervention by border control services in illegal activities, such as smuggling operations and piracy; routine surveillance and SAR missions conducted by coast guard services; as well as emergency services and first responders in disaster relief operations, for instance, after a major earthquake with a devastating tsunami or a catastrophic oil spill. Marine safety & security is critical for Canada to address the vulnerability of its sea lanes, ports and harbors to a variety of threats and illegal activities. Scarce surveillance and tracking capabilities make it difficult to keep track of all marine traffic across the length of Canada's coastline, which totals over 243,000 kilometres [1]. A coordinated response to emergency situations, like many safety and security operations, frequently involves a number of mobile agents, cooperatively performing missions and reacting to events that are distributed in time and space.

Situation Analysis (SA) is viewed as a process to provide and maintain a state of situation awareness for the decision maker [2]. Situation awareness is essential for decision-making activities: it concerns our perception of the elements in the environment, the comprehension of their meaning, and the projection of their status in the near future [3]. Computational models of situation analysis processes are in many cases of a distributed nature, being composed of multiple autonomously operating agents that each react in an asynchronous manner to discrete events distributed in space and time. Agents cooperate in developing a global understanding of a situation as it unfolds by exchanging information related to their local perception of events. Computer models support risk assessment and disaster response planning in the study of emergency preparedness by providing a practical alternative to real-world experiments, which in many cases are so costly and disruptive when performed on real situations that they do not provide a viable option.

This paper addresses the problem of testing complex computer models of infrastructure protection and emergency response situations based on detailed and realistic application scenarios in order to analyze and validate such models by means of advanced computational methods and tools. Although the primary focus in this paper is on marine safety & security operations [4, 5], we contend that the same approach carries over to a much broader scope of application fields. This is due to two facts: *1)* the concepts proposed here build on general abstractions of distributed systems composed of multiple mobile agents; and *2)* the way use cases are formally specified as abstract models of *vignettes* [6] in terms of *state machines* [7].

Methodical approaches to analyzing and testing decision support systems, situation analysis methods, and information fusion algorithms require realistic vignettes describing in great detail how use cases unfold depending on initial conditions for the configuration of agents, environmental aspects, and operational aspects. For instance, this includes the type and placement of agents in the geographic environment, their physical and operational capabilities, and their anticipated trajectories; specifics about environmental conditions, including weather, daylight, and water currents; and background noise, such as the number and distribution of neutral entities like any unrelated marine traffic. Realistic scenarios may include dozens or even hundreds of agents and complex conditions to be considered for each individual instance of a given vignette. Generating vignettes is a complex activity, one that is utterly inefficient to do manually and thus benefits greatly from automatic or semi-automatic approaches with tool support. We present here the conceptual design of a vignette generator to address this problem.

This paper is structured as follows. Section 2 outlines the problem scope, explains fundamental concepts and also the main goals of generating vignettes. Section 3 explains the compositional framework proposed for generating and/or composing vignettes out of reusable vignette elements. Section 4 discusses the conceptual design, including the main requirements, and the architecture of the system. Section 5 illustrates the underlying formalization approach for vignettes based on the *abstract state machine* method by means of common examples. Section 6 discusses applications and benefits of the Vignette Generator. Section 7 concludes the paper.

2 FUNDAMENTAL CONCEPTS

This section introduces the problem scope, some basic terminology, vignette patterns, and fundamental concepts used for describing the Vignette Generator in subsequent sections.

2.1 Problem Scope

Detailed experimental studies of emergency scenarios by means of computer simulation and animation play a crucial role in the development of innovative ICT solutions for situation analysis and decision support [6, 8–10]. Furthermore, they are gaining momentum as a viable alternative to performing experiments in a real-world setting because of the inevitable limitations of any large-scale real-world experiment. Scenarios explored in the work presented here deal with operations performed by Marine Security Operation Centres (MSOC) [11]. This includes SAR operations for passengers and vessels in distress, as well as the protection of sea lanes, ports and harbors against threats and illegal activities. The increasing volume of marine traffic [12] calls for advanced computer-based systems to support MSOC personnel in their daily missions by automating routine coordination tasks, and building on common surveillance technologies such as the Automated Identification System (AIS) [13]. For this purpose, it is essential to analyze scenarios, evaluate algorithms and assess the quality of solutions systematically by performing in-depth experimental studies based on realistic and meaningful test cases with a degree of detail beyond what could be done manually. The approach for the generation of test cases proposed here assumes interactive development, analysis and validation of complex vignettes for testing realistic scenarios.

2.2 Basic Terminology

We define the following basic terms to disambiguate the concepts of domain, scenario, vignette, vignette specification, and vignette element [6].

Domain Model

A *domain model* is a conceptual model of a given problem domain, defining the various types of entities that are relevant for this domain, their attributes, relationships, constraints, and behaviors.

Scenario

A *scenario* is a specific interpretation of a domain model for a given geographical area and time period. For instance, a marine safety & security scenario may refer to the Straight of Georgia between Vancouver Island and the mainland of British Columbia in the current time.

Vignette

A *vignette*, denoted by V in this paper, is a story embedded in a scenario. The story unfolds as a set of discrete events involving agents and the physical environment in which they operate, describing the distribution of agents and events in time and space. For instance, a smuggling operation in the Straight of Georgia, off the northern shore of Vancouver Island, can be described as a vignette embedded in a marine safety & security scenario.

Vignette Specification

A *vignette specification* is a precise and structured, static, text-based description of a vignette.

Vignette Element

A *vignette element*, denoted by E in this paper, is any identifiable part of a vignette specification.

Vignette Generator

The term *vignette generator* refers to a tool for generating vignette specifications based on vignettes.

A vignette specification generated by a vignette generator conceptually represents a *state machine model* that can be interpreted by a simulation environment to produce simulation runs.

2.3 Vignette Patterns

Vignette elements are basic building blocks for constructing vignettes and therefore vignette specifications. They usually have parameters that allow adjustment for a given target context. Vignette elements can be considered as *reusable patterns* in the context of marine safety & security. These four patterns serve as examples:

- *White Traffic Area Pattern*: In this pattern, there are a number of neutral agents (e.g., Fishing Boats) that are moving randomly within a specified area. Such agents are considered part of the background, and referred to as "white traffic" (see Figure 1(a)).

- *Routine Patrolling Pattern*: In this pattern, an agent with the ability to visually capture the environment (e.g., a Helicopter or UAV) moves within a predefined path and provides environmental information for other agents (see Figure 1(b)).

- *Rescuing Pattern*: In this pattern, there is an agent (e.g., a Fishing Boat) in a distress situation. Coast guard rescue team (e.g., a Helicopter) tries to rescue the people involved and also secure the trapped agent (see Figure 1(c)).

- *Rendezvousing Pattern*: This pattern involves a larger vessel, like a Cargo Ship, and a number of smaller boats, like Zodiacs. The bigger vessel stays at a fixed location (the Rendezvousing Point), while the smaller vessels go back and forth between the rendezvousing point and possibly a location close to the shoreline (the Beach Point). Figure 1(d) illustrates a rendezvousing pattern for a cargo ship and two zodiacs.

- Any combination of these vignette elements can be considered for generating a composite target vignette. Figure 1(e) is an example of a combination of all four vignette elements.

2.4 The Vignette Generator

As already mentioned, the Vignette Generator is a tool for developing well-defined vignettes and systematic generation of vignette specifications. Arguably, using a vignette generator has a number of decisive advantages as it makes vignette specification much *easier*, *more time efficient*, and *less error prone*. That is, users will be freed from the complicated and tedious task of describing low-level details and repeating routine activities. In particular, it allows the automatic generation of statistically relevant test cases as variations of vignette specifications that can be derived from a convenient high-level specification. The following simple example illustrates the role of the Vignette Generator for defining the White Traffic Area Pattern.

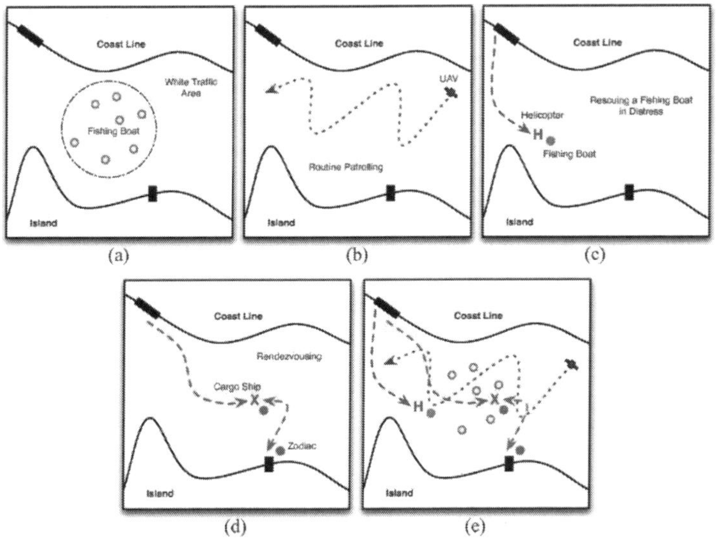

Figure 1 . (a) White-Traffic Area Pattern, (b) Routine Patrolling Pattern, (c) Rescuing Pattern, (d) Rendezvousing Pattern, (e) Composite Vignette.

Example 1

We need to specify 100 Fishing Boats within a given area, each equipped with a Communication Device with certain properties. A traditional approach, would have to define each Fishing Boat separately as follows:

```
WhiteTrafficArea
 FishingBoat-1
   Position(X, Y) = (x-1, y-1)
   CommunicationDevice-1
      Type = link-11
      Range = 5000 m
      .
      .
      .
 FishingBoat-100
   Position(X, Y) = (x-100, y-100)
   CommunicationDevice-100
      Type = link-11
      Range = 3000 m
```

In our proposed approach and tool, the vignette element (White Traffic Area Pattern) is described as follows:

```
WhiteTrafficArea
  #number = 100
  FishingBoat
     Position(X,Y) = #random value within area A1
     CommunicationDevice
        Type = link-11
        Range = #random value between (2000 m) and (5000 m)

  #: These elements and values are processed by the Vignette Gen-
erator engine.
```

Advantages

In the first specification, the details are constructed manually. In the second, the Vignette Generator will produce these details automatically. From the example, it is clear that the proposed approach has the following benefits.

1 Defining and understanding the second specification (7 lines) is much easier than the first one (501 lines).

2 Generating vignettes (or vignette elements) is more time efficient, specifically when the user needs to generate a number of vignette specifications based on a common pattern. This helps with statistical tests, which are necessary for evaluating algorithms.

3 Any error in the second specification is much easier to find and fix than in the first specification.

3. VIGNETTE COMPOSITION FRAMEWORK

In this section, we propose a framework for composing a new *Vignette Element Type* (ET) from a set of existing vignette element types.

Initial Type Set (I) is defined as a set of initial types which are the fundamental building blocks for composing new ETs. Although I can be extended if required, its suggested elements are as follows:

$$\left\{\begin{matrix} Angle,\ Boolean,\ DateTime,\ DeltaTime,\ Energy,\ Length, \\ Real,\ Speed,\ String,\ Temperature,\ Volume,\ Weight \end{matrix}\right\} \in \mathcal{I}$$

Vignette Element Type (ET) is considered as a building block for generating a *vignette* or another ET. Each ET has *name*, *type*, and also a set of ET s as its *attributes*.

Example 2

Before explaining our approach in detail, we illustrate the idea by means of a simplified ET, the *CH-149 Helicopter*. *CH-149* is a helicopter which has the *attributes* listed below. For instance, *OpticalSensor* ET is one of its attributes that in turn identifies specific attributes for this sensor element.

```
CH-149: Helicopter
  Height: Length
  Width: Length
  Length: Length
FuelCapacity: Volume
CruisingVelocity: Speed
MaximumVelocity: Speed
Communication: CommunicationDevice
   Frequency: Real
   Range: Length
   Bandwidth: Real
OpticalSensor: SensorDevice
   SamplingPeriod: DeltaTime
   Range: DeltaTime
```

A generated composite ET can be stored in a repository named *Vignette Repository* (R) for future use. Figure 2 illustrates a simple example of R.

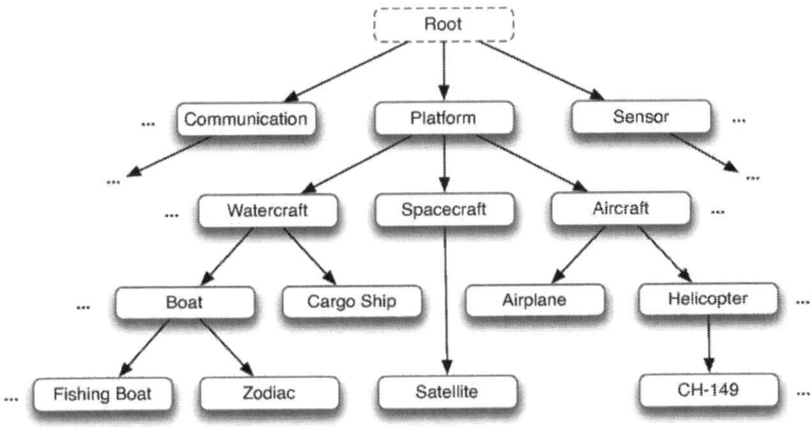

Figure 2. Example of the Element Repository (R).

*Vignette Element Types (*ET *) and Vignette Elements (*E*)*: We have defined Vignette Elements as identifiable parts of a vignette or vignette specification; also, they are considered as *instances* of Vignette Element Types. Thus, in order to include a Vignette Element in a vignette, we have to *instantiate* the appropriate ET , and then *assemble* it on the target *vignette* (see Section 3.1 for details).

3.1 Composition Mechanisms

Composition mechanisms are proposed with the purpose of creating new ET s and Es. These mechanisms are categorized in two different sets. The first set, ET-related mechanisms, is used to generate new composite Vignette Element Types from existing ETs in R. The second set, ET/E-related mechanisms, is used to generate Vignette Elements from appropriate ETs.

- ET *Mechanisms*

- *Extending*: Adding a Vignette Element Type *et'* to an existing element type *et* as its attribute.
- *Initiating*: Creating a new Vignette Element Type *et'* as a sub-type of another element type *et*.
- *Cloning*: Cloning an existing Vignette Element Type *et* with a different name *et'*.
- *Tuning*: Adjusting the value of the attributes of an existing ET□ .
- ET /E *Mechanisms*
- *Instantiating*: Creating an E as an instance of ET, and *tuning* all of its attributes.
- *Assembling*: Adding an instantiated E to the target vignette V , and therefore vignette specification.

All in all, a Vignette V is a set of Vignette Elements Es which are instances of Vignette Element Types ET s (see Figure 3).

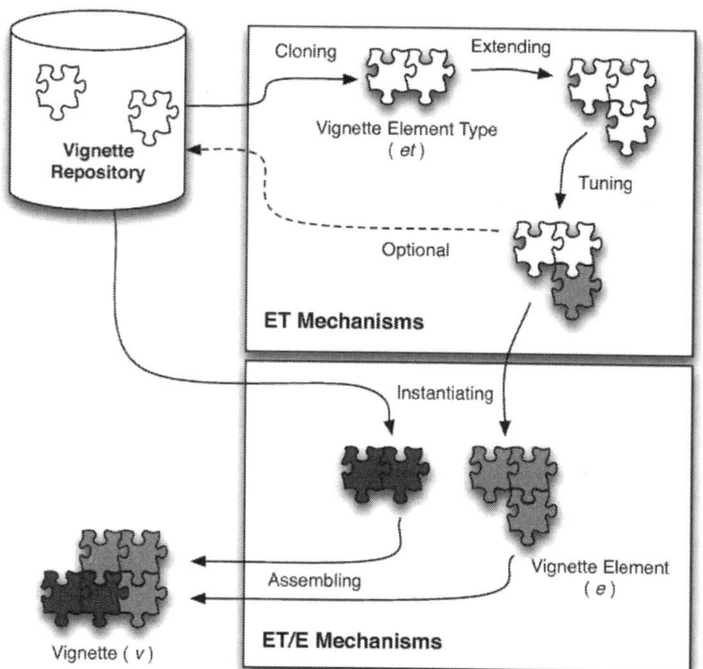

Figure 3 . Example of a Hybrid Composition Mechanism.

3.2 Relationships between Vignette Element Types

Two different types of relationships are defined among ET s:

- *Has-a Relationship*: Each ET can have a number of ET s as its *attributes*.
- *Is-a Relationship*: ET s can have hierarchical relationships. In this way, each ET has a super-type (except for the root element). All sub-types of an ET inherit all *has-a* relationships from their super-type. However, they are able to override these inherited relationships. For example, if an ET (et_1) *has-a* ET (et_2), all sub-types of et_1 will have et_2. By this approach, we prevent redundancies in defining ETs and their relationships. Moreover, it helps to maintain the ET s in a more efficient way.

Since the number of Vignette Element Types can increase considerably over time, these relationships help in organizing them within the repository.

Example 3

The composition of a Routine Patrolling Pattern (Figure 1(b)) and Rendezvousing Pattern (Figure 1(d)) is illustrated here. By applying the following steps, we can generate the composed vignette. Assume that we already have *CargoShip*, *Zodiac*, and *UAV* ET s in the repository.

1 *Initiate* Routine Patrolling Pattern ET as a subtype of *root* ET

2 *Extend* Routine Patrolling Pattern ET with a *UAV* ET

3 *Initiate* Rendezvousing Pattern ET as a subtype of *root* ET

4 *Extend* Rendezvousing Pattern ET with a *CargoShip* ET

5 *Extend* Rendezvousing Pattern ET with two *Zodiac* ET

6 *Tune* the *trajectory* attributes of the *CargoShip* and *Zodiac* ET

7 *Instantiate* Routine Patrolling Pattern ET and *Tune* all of its attributes to generate Routine Patrolling Pattern E

8 *Instantiate* Rendezvousing Pattern ET and *Tune* all of its attributes to generate Rendezvousing Pattern E

9 *Assemble* Routine Patrolling Pattern E and Rendezvousing Pattern E into a new current vignette (V

4. CONCEPTUAL DESIGN

This section discusses the conceptual design of the Vignette Generator, including the system requirements and high-level architecture.

4.1 System Requirements

The requirements of the Vignette Generator outlined here were elicited and analyzed in part through discussion with domain experts. Afterward, high level requirements were identified and classified into different categories. For each category, the most significant lower level requirements were explored and discussed.

- *Flexible support for a variety of vignettes*: The Vignette Generator should support definition of different vignette elements, with fixed or variable characteristics, that can be reused and combined in creating vignette specifications.
  - The ability to easily specify a set of vignette elements implicitly by specifying a vignette element together with a set of parameters. The Vignette Generator should then automatically generate an explicit set.
    - Example: In the White Traffic Area Pattern (Figure 1(a)), the user can specify a single concrete Fishing Boat (an existing vignette element from the repository), and set a few parameters such as the number of boats desired and an area. Then the Vignette Generator produces a specification with that many boats distributed over this area, moving about in a certain manner. The parameter varied is the initial trajectory of each boat (see Example 1).
  - The ability to aggregate, name, store, and retrieve previously defined vignette elements in a systematic way.

    - Example: A new vignette element with some specific features and attributes can be given a name, e.g., CH-149 Helicopter, and stored in the repository. Later on, we should be able to reuse the stored element as is or extend it for other uses.

    - Example: The White Traffic Area Pattern and the Rendezvousing Pattern (Figure 1) can be stored in the repository, and reused to compose a new complex vignette.

  - The ability to expand the repository of vignette elements over time which leads to the gradual generation of more and more complex vignette elements and vignettes.

- *Support for non-determinism*: The user should be able to abstract away from the detailed configuration of vignette elements through non-deterministic value assignments to the attributes of vignette elements.

  - The ability to easily specify a set of vignette elements from which the generator picks one at random.

    - Example: If Aircraft is specified, the generator randomly substitutes one of the concrete platforms that is a subtype of Aircraft, such as the CH-149 Helicopter (Figure 2).

    - Example: In a traffic area that we have a number of platforms which have communication devices, the generator randomly assigns one of the concrete communication devices that is a subtype of Communication (Figure 2).

  - The ability to easily specify parameterized (unfixed) values (e.g., randomly, from a range or distribution) for attributes of vignette elements.

    - Example: The value of Velocity attribute of CH-149 Helicopter should be selected randomly from a range between X and Y ($X \geq Velocity \geq Y$).

- *Validation of high-level vignettes*: The Vignette Generator should allow validation of scenario descriptions against domain-specified and user-defined sets of constraints.
  - The ability to automatically check for errors, inconsistencies, policy and best-practice violations, and omissions.
    - Example: The generator should issue warnings for integrity violations such as inconsistent behavior, e.g., ships traveling over land or the use of communication equipment that is incompatible with all others.
  - The ability to automatically check for user-defined constraints. The Vignette Generator should issue warnings for any violations.
    - Example: The value of the Altitude attribute of all Aircraft should be between a minimum and a maximum value; otherwise, the system should report a warning. ($X \geq Altitude \geq Y$).
- *Production of concrete vignette specifications for different simulation environments*: From every abstract vignette that complies with the requirements of a given simulation environment, the Vignette Generator should be able to generate a number of concrete vignette specifications in a specified format (e.g., XML) that can be loaded in the simulation environment. To this end, the engine of the Vignette Generator should not depend on any specific simulation environment. This requirement is useful for generating vignettes for different simulators or even a simulator under construction.
- *Easy to use Graphical User Interface (GUI)*: The Vignette Generator should come with an easy to use GUI, which supports the creation, modification, and instantiation of vignette elements.

4.2 System Architecture

To address the mentioned functional and non-functional requirements, we propose a two-layer architecture (see Figure 4).

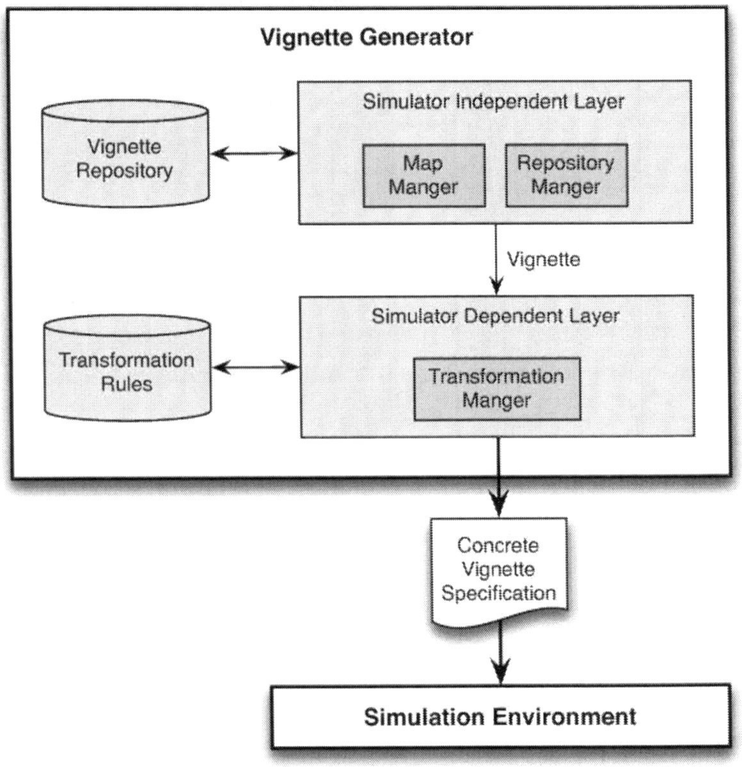

Figure 4 . Vignette Generator Architecture.

- *Simulator Independent Layer*: This layer is responsible for composing a new vignette based on Vignette Element Types and Vignette Elements. The output is not executable and needs to be transformed into the required format. The two main components of this layer are the *map manager* and the *repository manager*.

- *Simulator Dependent Layer*: This layer transforms the output of the *simulator independent layer* into a format that can be processed and executed by the simulation environment. The main component of this layer is the *transformation manager*.

- *Vignette Repository*: The *vignette repository* is a knowledge base for storing all information about the ET s, their attributes and relationships.

- *Transformation Rules*: Transformation rules which are used by the *simulator dependent layer* are stored here.

5 FORMAL REPRESENTATION

Based on common concepts of computational logic and discrete mathematics, we formalize vignette specifications as *abstract state machine* (ASM) models [7, 14]. Abstract state machines are known for their versatility in modeling semantic properties of algorithms, architectures, languages, protocols, and virtually all kinds of sequential, parallel and distributed systems. Building on an abstract machine framework has a number of advantages for defining structural and behavioral properties of vignettes, and the composition of simple and complex vignette elements so as to form complex structures.

5.1 Abstract State Machines and CoreASM Tool Environment

This section briefly presents the basic modeling concepts of Abstract State Machines and the CoreASM modeling framework for simulation and analysis of ASM specification.

5.1.1 Abstract State Machines

The ASM method [7] is a versatile mathematical framework for modeling virtually all kinds of discrete dynamic systems, including sequential, parallel and distributed systems, at any desired level of abstraction with a noticeable orientation toward practical applications. Building on common and widely used concepts from discrete mathematics and computational logic, it combines *abstract states* with *transition systems*. Abstract state machines are known for their semantic foundations for architectures, languages, and protocols, including some of the most prominent ones such as Java, SDL, and VHDL [15–20].

We use here an asynchronous ASM computation model, i.e., Distributed Abstract State Machine (DASM), with a non-empty set of autonomously operating computational *agents*. This set can change dynamically over machine runs so as to model a varying number of computational resources. The asynchronous computation model of DASM defines concurrent and reactive behavior, as observable in distributed computations performed by autonomously operating computational agents, according to the underlying semantic model. Depending on the agent type, agents have different dynamic properties as defined by their ASM program. Agents interact with each other and their operational environment by reading and writing shared locations of global machine *states*, represented as first-order structures [21] in terms of sets and operations defined thereon. Agents access and manipulate machine states as described by the transition rules that form their program. The underlying semantic model resolves potential conflicts according to the definition of *partially ordered runs* [7].

5.1.2 The CoreASM Extensible Architecture

CoreASM [14, 22] is an Open Source project and tool suite for rapid proto-typing, analysis and experimental validation of ASM models. The tool suite provides a platform-independent execution engine and a GUI for interactive visualization and control of simulation runs. Building on an extensible plugin-based architecture [23], the CoreASM kernel (the *core* of the language and engine) contains only bare essentials; most of the language constructs and functionalities of the engine come in the form of plugins extending the kernel [24].

There are two flexible mechanisms for extending CoreASM. Plugins can either extend the functionality of specific engine components (such as the parser or the scheduler), introducing additional data or behavior to those components, or they can extend the control flow of the engine by interposing their own code in between state transitions of the engine. This extensibility offers a great deal of flexibility for customizing CoreASM depending on specific application needs.

Over several years, CoreASM has been put to the test in a range of R&D applications in commercial enterprises and government agencies, spanning computational criminology, coastal surveillance, SA, decision support systems, and Web services. CoreASM is implemented in Java under Academic Free License version 3.0 (AFL 3.0), providing a sensible compromise between public availability of the original source code and the existence of proprietary extensions for commercial applications. It is readily available at http://www.coreasm.org.

5.2 ASM Models of Vignettes

Intuitively, a vignette corresponds to an abstract machine *program*, and a vignette execution corresponds to an abstract machine *run*. Specifically, we use the asynchronous computation model of distributed abstract state machines, defined in terms of autonomously operating computational agents interacting with each other and their operational environment (the external world) by reading and writing locations of globally shared states. Potential conflicts are resolved by the underlying semantic model of partially ordered runs (see [7] for details).

Figure 5 illustrates a sample ASM model for a Rendezvousing Pattern between two agents at a high level of abstraction. The rendezvousing operation described here assumes that agent A_1 is a larger stationary vessel (for instance, a cargo ship), while A_2 is a smaller boat going back and forth between A_1 and a beach location to load and unload smuggled goods. The operation Rendezvousing(..) specifies the decomposition of a vignette into its constituent elements, expressing how A_1 and A_2 move to the off-shore rendezvousing point and start the loading and unloading process. Exact coordinates for the rendezvous location and the beach point, as well as the detailed trajectory of A_2, are to be determined at runtime by the simulator executing the vignette.

A sample ASM model for a Rescuing Pattern among three agents is shown in Figure 6. In this SAR operation, a helicopter agent A_2 first searches and then identifies the location of the fishing boat agent A_1 in a distress situation. After locating the fishing boat, a coast guard vessel agent A_3 first extracts persons from the fishing boat and then secures the boat. The operation MoveAgent(..) is defined in the Rendezvousing example.

```
AGENT ≡ {CARGOSHIP, ZODIAC, ...}
MODE ≡ {APPROACHING, SMUGGLING, ...}
SMUGGLINGMODE ≡ {LOADING, UNLOADING, ...}
universe POSITION    // Geographic coordinates
universe OPERATION   // Operations of a vignette

AgentPosition : AGENT ⇸ POSITION
Mode : OPERATION ⇸ MODE
SmugglingMode : OPERATION ⇸ SMUGGLINGMODE
BeachPoint : OPERATION ⇸ POSITION
RendezvousPoint : OPERATION ⇸ POSITION

Rendezvousing(A1, A2 : AGENT; Op : OPERATION) ≡
  if Mode(Op) = APPROACHING then
    MoveAgent(A1, RendezvousPoint(Op))
  if Mode(Op) = SMUGGLING then
    if SmugglingMode(Op) = LOADING then
      MoveAgent(A, RendezvousPoint(Op))
    if SmugglingMode(Op) = UNLOADING then
      MoveAgent(A, BeachPoint(Op))

MoveAgent(A : AGENT; P : POSITION) ≡
  NextPosition(A, P), ...
  // Incrementally generates a trajectory by computing intermediate agents positions
  // depending on various factors, gradually moving an agent towards its destination.
```

Figure 5 . ASM model of a Rendezvous Pattern between two agents **A** 1 , **A** 2 in Territorial Sea (a belt of coastal waters extending from the baseline of a coastal state): **A** 1 is of type C **ARGO** S **HIP** and **A** 2 is of type Z ODIAC.

6 APPLICATIONS

An operational prototype of the Vignette Generator has been designed and developed in Java based on the concepts presented in this paper. It is capable of generating a wide spectrum of vignette specifications for the INFORM Lab simulation environment (see Section 6.1.2). A screenshot of the Vignette Generator illustrating a complex SAR scenario is illustrated in Figure 7.

```
AGENT ≡ {FISHINGBOAT, HELICOPTER, COASTGUARDBOAT...}
MODE ≡ {SEARCHING, RESCUING, SECURING, ...}
universe POSITION     // Geographic coordinates
universe AREA     // Geographic areas
universe OPERATION     // Operations of a vignette
AgentPosition : AGENT ⇸ POSITION
Mode : OPERATION ⇸ MODE
Area : OPERATION ⇸ AREA
Rescuing(A1, A2, A3 : AGENT; Op : OPERATION) ≡
  if AgentPosition(A1) = undef then
    if Mode(Op) = SEARCHING then
      SearchArea(A1, A2, Area(Op))
      // Searches for A1 in a specified area and sets the position of A1 once identified.
  if AgentPosition(A1) ≠ undef then
    if Mode(Op) = RESCUING then
      RescuePersons(A1, A3)
    else if Mode(Op) = SECURING then
      SecureBoat(A1, A3)
RescuePersons(A1, A3 : AGENT) ≡
  if AgentPosition(A1) ≠ AgentPosition(A3) then
    MoveAgent(A3, AgentPosition(A1))
  else
    PerformRescue(A1, A3)
SecureBoat(A1, A3 : AGENT) ≡
  if AgentPosition(A1) ≠ AgentPosition(A3) then
    MoveAgent(A3, AgentPosition(A1))
  else
    PerformSecure(A1, A3)
```

Figure 6 . ASM model of Rescuing Pattern among three agents in Territorial Sea: Agent A_1 is of type F ISHING **B** OAT **,** A_2 is of type H ELICOPTER **,** and A_3 is of type C OAST **G** UARD **B** OAT.

The Vignette Generator has the potential to interoperate with different simulation environments, namely those for which the corresponding interface configuration format is determined, due to the following features:

- *Composable Vignette Element Type*: Vignette V is generated from ETs forming the basic building blocks. So, new ETs can be generated based on existing ones at run-time, which means that R is *flexible* enough that users can easily *extend* the set of ETs, and consequently R□ .

- *Separated Transformation Layer at the Architecture Level*: Generating a vignette Vfrom high-level abstract ETs, and transforming V into an executable format (for the simulation environment) are separated into two different architectural layers. The input format of the simulation environment is irrelevant for the composition of ETs and E s. This way, the two-layer architecture enhances of the *portability* of the Vignette Generator.

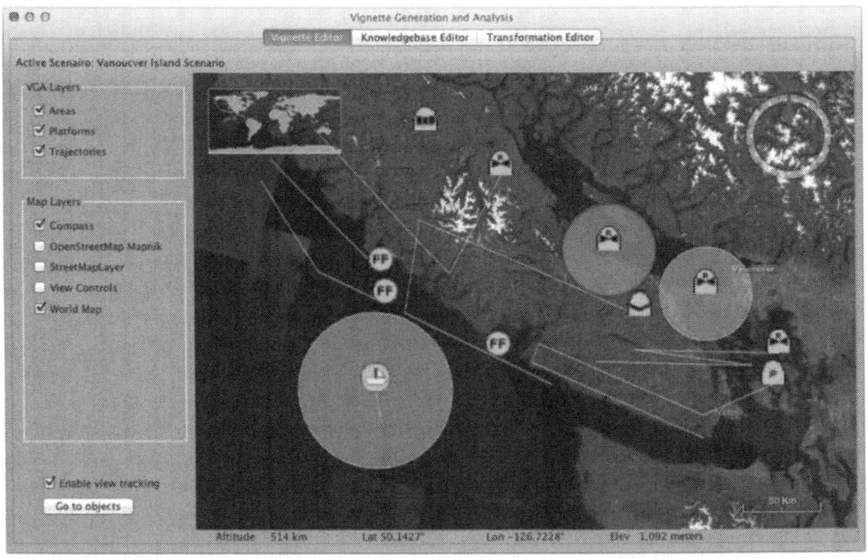

Figure 7. A Screenshot of a Map-Based Scenario Representation in the Vignette Generator.

Therefore, we are able to generate various vignettes by instantiating ET s with less consideration of the input format required by a specific simulation environment. We have designed the Vignette Generator such that it can provide different simulation environments with appropriate inputs without significant modifications. In the following section, realistic applications of the tool are discussed.

6.1 Generating Application Domain Vignettes

A detailed sample scenario, including a number of related vignettes, serves to illustrate the functionality of the proposed Vignette Generator. The generated vignettes, which can be executed in the INFORM Lab simulation environment, are based on the sample scenario presented in Section 6.1.1. In this scenario two different kinds of vignettes are presented: cooperative search, such as locating a fishing boat in distress, and non-cooperative search to detect and prevent illegal activities (for instance, smuggling operations).

6.1.1 Sample Scenario and Vignettes

Consider a simulation of maritime activities near Vancouver Island, B.C., Canada on December 26, 2011 between 6:30 AM and 7:30 AM. These maritime activities include the following events and stories:

- December 26, 2011 - 6:30 AM:
- Routine patrolling of Vancouver Island's coastal region by *coast guard vessels* and *helicopters*.
- Monitoring the maritime environment through *satellite* images.
- December 26, 2011 - 6:45 AM:
- Receiving a rescue request from a *fishing boat* in a distress situation with an approximate reported location southeast of the Gulf Islands in the Strait of Georgia between Vancouver Island and the mainland of British Columbia.
- December 26, 2011 - 6:50 AM:
- Observing suspicious offshore activities between a *cargo ship* and one or more *small boats (zodiacs)* roughly 50 kilometres northwest of Port Hardy at the northern end of Vancouver Island.
- December 26, 2011 - 7:00 AM:
- Searching and rescuing the passengers of the *fishing boat* and securing their boat as a cooperative SAR mission.

- December 26, 2011 - 7:10 AM:
- Analyzing the suspicious offshore activities by a *coast guard helicopter* confirms the suspicion of a smuggling operation in progress.
- In addition to these vignettes, there are also a number of *fishing boats*, *sailing boats*, *cargo vessels*, *passenger vessels*, and *tankers* operating in the area of interest. Since one of the main purposes of the Vignette Generator is to test different situation analysis methods and information fusion algorithms, considering such "background traffic" is necessary to produce realistic noise, say for testing offshore smuggling operation detection algorithms under realistic conditions.

The following four vignettes are elicited from these stories:

- Routine patrolling and monitoring of the coastal regions of Vancouver Island by *coast guard vessels*, *helicopters*, and *satellites* (see Figure 8(a)).
- Searching and rescuing a *fishing boat* in a distress situation by *coast guard vessels* and *SAR helicopters* (see Figure 8(b)).
- Observing and detecting suspicious activities between a *cargo ship* and *two zodiacs* (see Figure 8(c)).
- Random movements of various types of smaller boats and larger vessels in a maritime environment (see Figure 8(d)).

Combinations of the above high-level vignettes (instantiated with a number of nondeterministically chosen parameter values) can be defined in the Vignette Generator. As a result, one can easily generate a number of concrete instances of these vignettes as input for the INFORM Lab simulation environment. Figure 9 presents a screenshot from the INFORM Lab simulation of a concrete instance of the vignette shown in Figure 7.

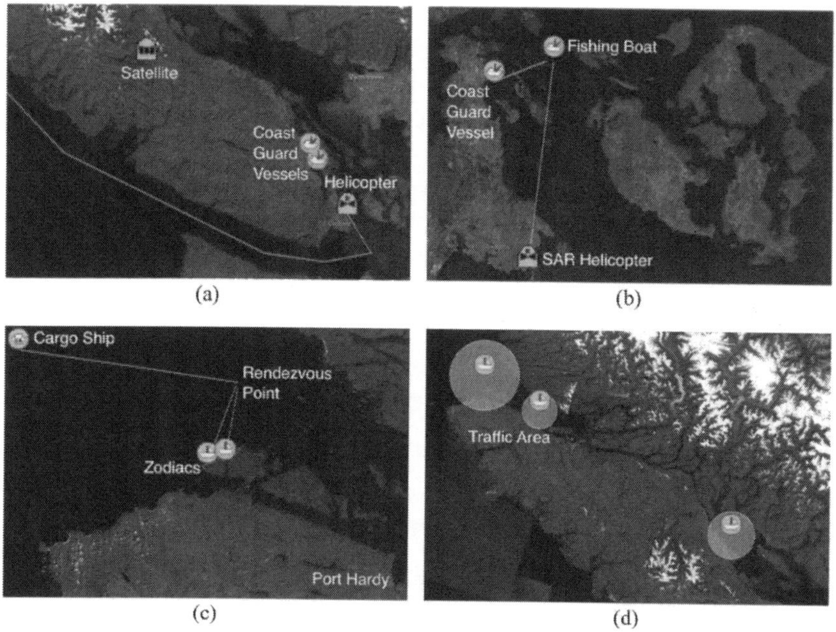

Figure 8 . Sample Vignettes from the Vignette Generator (a) Routine Patrolling: Coast guard vessels, helicopters, and satellite images are used to patrol and monitor the area of interest; (b) Search and Rescue: A fishing boat in a distress situation is located southeast of the Gulf Islands in the Strait of Georgia, and a coast guard vessel and a helicopter are performing SAR operation; (c) Rendezvousing: Suspicious offshore activities between a cargo ship and two zodiacs are observed roughly 50 kilometres northwest of Port Hardy; (d) Traffic Area: There are a number different kinds of vessels (including fishing boats, sailing boats, passenger vessels, cargo vessels, and tankers) operating in the area of interest.

The examples described here are well founded in real-world situations and have been examined thoroughly and validated by domain experts. However, it is important to note that they describe a situation at a very local scale. More complex vignettes can involve hundreds of agents and platforms, as well as numerous distinct behaviors. The Vignette Generator is fully capable of handling vignettes of high complexity and large scale, but a full description of such a comprehensive example would be outside of the scope of a journal article.

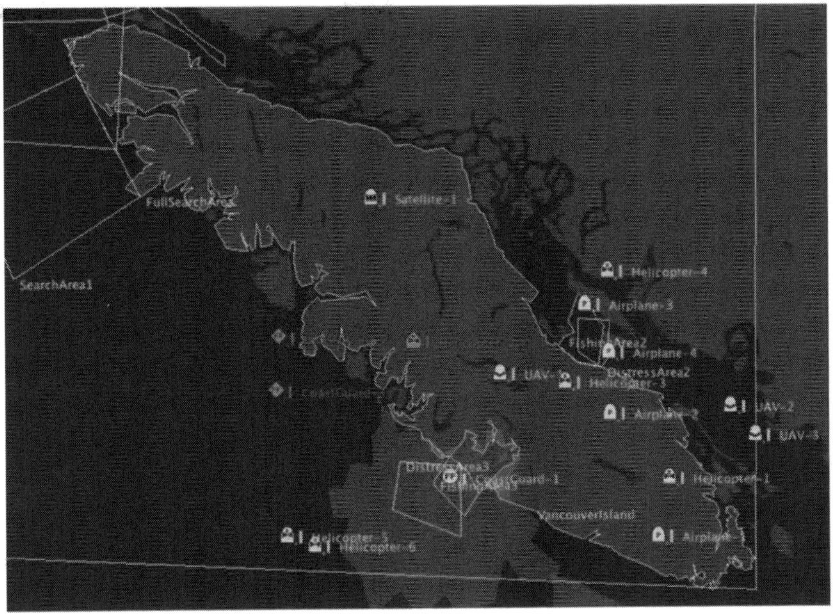

Figure 9 A Screenshot of the INFORM Lab's Simulator: Detection of a Smuggling Operation in the Strait of Georgia at the northern peak of Vancouver Island.

6.1.2 INFORM Lab Simulation Environment

INFORM Lab, designed and developed at MDA Corporation, Richmond, B.C., Canada, is an advanced agent based simulation framework built around the OODA paradigm (Observe, Orient, Decide, Act) [8]. It is used to simulate Coastal Wide Area Surveillance applications, including SAR and detection of illegal activities (see Figure 9). It allows experimentation with higher-level distributed dynamic information fusion, distributed dynamic resource management, communication strategies, and configuration management given multiple constraints on resources and their communications networks. So, it can be considered as a testbed which also allows the evaluation of sensing strategies, motion strategies, and a range of control strategies from independent agent operation through various levels of agent collaboration.

The INFORM Lab simulator is driven by a configuration specification file that specifies the agents and their relationships, the environment, initial conditions as well as any scripted events. Setting up this file can be time-consuming, difficult and error prone. However, being a testbed, INFORM Lab requires that statistically relevant metrics can be extracted from the results. This means that many small variations of an experiment need to be run. Specifying this in a convenient, systematic, and error-free manner would be a huge task without a vignette generator. The Vignette Generator can provide the required variability of the experiments with great convenience and is time-saving for the experimenter.

6.2 Anomaly Detection

Related to the work presented here is a research project on a model-driven framework for engineering Situation Analysis Decision Support (SADS) systems for the domain of marine safety & security [25, 26]. SADS system engineering practices call for systematic formal modeling approaches to manage complexity through modularization, refinement and validation of abstract models. In this light, we explore SADS system design based on ASM modeling techniques paired with CoreASM tool support to facilitate analysis of the problem space and reasoning about design decisions and conformance criteria so as to ensure they are properly established and well understood prior to building the system. We provide an extension to CoreASM for the marine safety & security domain, Specifically for capturing rendezvous scenarios. The extension yields the necessary background concepts, such as mobile sensors and shipping lanes, and offers runtime visualization of simulation runs together with an analyzer to measure success of various rendezvous detection strategies used in the model. The proposed framework complements purely analytical means that focus on verification of internal properties, such as consistency and completeness of a model, and provides a sensible way of linking formal and empirical aspects in the model-driven engineering of SADS systems. Experimental studies of SA scenarios can considerably enhance our insight into intricate system dynamics and simplify the challenging task of deriving meaningful conformance criteria for checking the validity of SADS domain models against established operational concepts of marine safety & security. We illustrate the application of the proposed approach using sample rendezvous scenarios (see [25] for a detailed scenario description and results). Work in progress is now extending the scope of our SADS model to capture a much wider range of observable behaviors of marine traffic that deviate from what is considered normal. This work aims at integrating model-driven approaches with data-driven approaches of behavior detection into a hybrid detection framework operating on real-world data sets.

6.3 General Benefits of Using the Vignette Generator

- *Ease of Use*: Graphical user interface facilitates manipulation of the repository and the transformation rules, and helps users to generate appropriate test cases.
- *Reusability*: By gradually building a repository of ET s, we are able to reuse the defined ET □ s in future applications.
- *Compositionality*: By using the proposed composition mechanisms, we are able to generate increasingly more complex ET s and Vs by combining and orchestrating existing ET □ s.
- *Complexity Management*: The hierarchical structure of the repository reduces the complexity of generating vignettes. Low-level details for an ET need only be defined once and can then be reused. Further, a composite ET hides the low-level details of its internal elements without preventing access to them. *Incremental Improvement*: Over time, generating complex vignettes becomes increasingly efficient as the repository grows, offering more and richer choices.

7 CONCLUSION AND FUTURE WORK

This paper addresses a notorious problem in testing computational models for infrastructure protection and emergency response by means of simulation and animation using realistic scenarios. Simulation plays a key role for analyzing decision support systems, situation analysis methods, and information fusion algorithms. Considering their inherent complexity, the validity of realistic models need to be established through progressive comprehension of the real-world phenomena being studied. This can be done by gradually improving the applied methods in an interactive manner. However, such improvements are virtually impossible without preforming in-depth experiments. Furthermore, manual generation of meaningful vignettes is tedious and cumbersome, and generating such vignettes even in an automatic or semi-automatic fashion is a complex and challenging engineering task.

We have presented the conceptual design for a vignette generator to overcome this problem, and described its high-level requirements and architecture in detail. Further, we have illustrated a formal approach to vignette specification in terms of a sample vignette, building on well-defined composition mechanisms. As justified in the paper, the proposed design ensures basic quality attributes, including *flexibility*, *extensibility*, *reusability*, *scalability*, and *portability*. Practical experience has been gained from using the Vignette Generator with the INFORM Lab simulation environment at MDA Corporation for developing situation analysis decision support models. Generating vignettes for marine safety & security scenarios undeniably helps developers to test and evaluate their algorithms. It also helps end-users to simulate and analyze real-world situations. Arguably, the scope of potential applications of the Vignette Generator presented here extends far beyond the domain of marine safety & security operations as many central concepts directly carry over to a much broader range of infrastructure protection and emergency response scenarios.

ACKNOWLEDGEMENTS

The work presented here has been funded by MDA Corporation, MITACS, and NSERC under NSERC Collaborative Research and Development Grant No. 342503-06.

COMPETING INTERESTS

The authors declare that they have no competing interests.

AUTHORS' CONTRIBUTIONS

This paper is the outcome of a joint effort with major contributions particularly from the first three authors. All authors have contributed to the paper, read and approved the final manuscript.

REFERENCES

1. Natural Resources Canada: Natural Resources Canada. The Atlas of Canada - Coastline and Shoreline [Online].2011. [http://atlas.nrcan.gc.ca/site/english/learningresources/facts/coastline.html]
2. Bossé É, Roy J, Ward S: Models and Tools for Information Fusion. Artech House; 2007.
3. Endsley MR: Theoretical Underpinnings of Situation Awareness: A Critical Review. In Situation Awareness Analysis and Measurement. Edited by: Endsley MR, Garland DJ. LEA; 2000:3–32.
4. Glässer U, Jackson P, Araghi AK, Shahir HY, Wehn H: A Collaborative Decision Support Model for Marine Safety and Security Operations. In Distributed, Parallel and Biologically Inspired Systems. Edited by: Hinchey M, et al. Springer; 2010.
5. Glässer U, Jackson P, Araghi AK, Shahir HY: Intelligent Decision Support for Marine Safety and Security Operations. In Proceedings of the 2010 IEEE International Conference on Intelligence and Security Informatics, Vancouver, Canada. Edited by: Yang C, et al. IEEE; 2010:101–107.
6. Lambert D, Wark S, Bossé É, Pigeon L, Blackman C, Hinma M: A Coalition Approach to Higher-Level Fusion. In Proceedings of the International Conference on Information Fusion, Seattle, USA. IEEE; 2009.

7. Börger E, Stärk R: Abstract State Machines: A Method for High-Level System Design and Analysis. Springer-Verlag; 2003.

8. Li Z, Leung H, Valin P, Wehn H: High Level data fusion system for CanCoastWatch. Proceedings of the 10th International Conference on Information Fusion 2007, 1–6.

9. Jackson P, Glässer U, Shahir HY, Wehn H: An Extensible Decision Engine for Marine Safety and Security. In Proceedings of the 2011 IEEE International Conference on Intelligence and Security Informatics, Beijing, China. IEEE; 2011:54–59.

10. Shahir HY, Glässer U, Jackson P, Wehn H: Test-Case Generation for Marine Safety and Security Scenarios. In Proceedings of the 2011 IEEE International Conference on Intelligence and Security Informatics, Beijing, China. IEEE; 2011:48–53.

11. Royal Canadian Mounted Police: Marine Security Operation Centres (MSOC) [Online].2011. [http://www.grc-rcmp.gc.ca/mari-port/msoc-cosm-eng.htm]

12. MarineTraffic Service: Vessel positions tracking based on AIS data.2011. [http://www.marinetraffic.com/]

13. U.S. Department of Homeland Security - U.S. Coast Guard: Automatic Identification System Overview.2011. [http://www.navcen.uscg.gov/?pageName=AIS]

14. Farahbod R, Gervasi V, Glässer U: CoreASM: An Extensible ASM Execution Engine. Fundamenta Informaticae 2007,77(1–2):71–103.

15. Glässer U, Gotzhein R, Prinz A: The Formal Semantics of SDL-2000: Status and Perspectives. Computer Networks 2003,42(3):343–358. 10.1016/S1389-1286(03)00247-0

16. Börger E, Glässer U, Müller W: Formal Definition of an Abstract VHDL'93 Simulator by EA-Machines. In Formal Semantics for VHDL. Edited by: Delgado Kloos C, Breuer PT. Kluwer Academic Publishers; 1995:107–139

17. Glässer U, Gurevich Y, Veanes M: Abstract Communication Model for Distributed Systems. IEEE Trans on Soft Eng 2004,30(7):458–472. 10.1109/TSE.2004.25

18. Stärk R, Schmid J, Börger E: Java and the Java Virtual Machine: Definition, Verication, Validation. Springer-Verlag; 2001.

19. Farahbod R, Glässer U, Vajihollahi M: An Abstract Machine Architecture for Web Service Based Business Process Management. International Journal of Business Process Integration and Management 2007, 1: 279–291.

20. Börger E, Riccobene E, Schmid J: Capturing Requirements by Abstract State Machines: The Light Control Case Study. Journal of Universal Computer Science 2000,6(7):597–620.

21. Enderton HB: A Mathematical Introduction to Logic. second edition. Harcourt Academic Press; 2001.

22. Farahbod R, Glässer U: The CoreASM Modeling Framework. Software: Practice and Experience 2011,41(2):167–178. 10.1002/spe.1029

23. Farahbod R, Gervasi V, Glässer U, Ma G: CoreASM Plug-in Architecture. In Rigorous Methods for Software Construction and Analysis. Edited by: Abrial JR, Glässer U. Springer; 2009:147–169. LNCS Festschrift volume 5115 LNCS Festschrift volume 5115

24. Farahbod R: CoreASM: An Extensible Modeling Framework & Tool Environment for High-level Design and Analysis of Distributed Systems. PhD thesis. Simon Fraser University, BC, Canada; 2009.

25. Farahbod R, Avram V, Glässer U, Guitouni A: Engineering Situation Analysis Decision Support Systems. Proceedings of the European Intelligence and Security Informatics Conference, Athens, Greece 2011, 10–18.

26. Farahbod R, Avram V, Glässer U, Guitouni A: A Formal Engineering Approach to High-level Design of Situation Analysis Decision Support Systems. Proceedings of the 13th International Conference on Formal Engineering Methods, Durham, UK 2011, 211–226.

CHAPTER 10

Radio Hazard Safety Assessment for Marine Ship Transmitters: Measurements Using a New Data Collection Method and Comparison with ICNIRP and ARPANSA Limits

Malka N. Halgamuge *

Department of Electrical and Electronic Engineering,
The University of Melbourne, Parkville, VIC 3010, Australia;

ABSTRACT

We investigated the levels of radio frequency electromagnetic fields (RF EMFs) emitted from marine ship transmitters. In this study, we recorded the radio frequency (RF) electric field (EF) levels emitted from transmitters from a marine vessel focusing on the areas normally occupied by crew members and passengers. Previous studies considered radiation hazard safety assessment for marine vessels with a limited number of transmitters, such as very high-frequency (VHF) transceivers, radar and communication transmitters. In our investigation, EF levels from seven radio transmitters were measured, including: VHF, medium frequency/high frequency (MF/HF), satellite communication (Sat-Com C), AISnavigation, radar X-band and radar S-band. Measurements were carried out in a 40 m-long, three-level ship (upper deck, bridge deck and bridge roof) at 12 different locations. We developed a new data-collection protocol and performed it under 11 different scenarios to observe and measure the radiation emissions from all of the transmitters. In total, 528 EF field measurements were collected and averaged over all three levels of the marine ship with RF transmitters: the measured electric fields were the lowest on the upper deck (0.82–0.86 V/m), the highest on the bridge roof (2.15–3.70 V/m) and in between on the bridge deck (0.47–1.15 V/m). The measured EF levels were then assessed for compliance with the occupational and general public reference levels of the International Commission on Non-Ionizing Radiation Protection (ICNIRP) guidelines and the Australian Radiation Protection and Nuclear Safety Agency (ARPANSA) standards. The ICNIRP and the ARPANSA limits for the general public were exceeded on the bridge roof; nevertheless, the occupational limits were respected everywhere. The measured EF levels, hence, complied with the ICNIRP guidelines and the ARPANSA standards. In this paper, we provide a new data collection model for future surveys, which could be conducted with larger samples to verify our observations. Furthermore, this new method could be useful as a reference for researchers and industry professionals without direct access to the necessary equipment.

KEYWORDS

EMF; electric fields; radio hazard; ICNIRP limits; marine shipboard transmitters; exposure of general public

1. INTRODUCTION

The International Commission on Non-Ionizing Radiation Protection (ICNIRP) is an independent scientific organization that offers regulations and recommendations about the health and environmental effects of non-ionizing radiation exposure [1]. The exposure limits recommended by international standards-setting bodies—ICNIRP [1], the Institute of Electrical and Electronics Engineers (IEEE) [2] and the European Committee for Electrotechnical Standardization (CENELEC) [3]—are based on short-term and immediate health effects of elevated tissue temperatures due to the absorption of energy during exposure to non-ionizing radiation. In 2002, the Australian Radiation Protection and Nuclear Safety Agency (ARPANSA) published a standard that specifies health limits of human exposure to radio frequency fields ranging from 3 kHz–300 GHz in order to avoid its adverse health effects [4]. These guidelines aim to minimise the heating effects caused by the absorption of radio frequency electromagnetic fields (RF-EMFs). Each of the standards indicate basic restrictions for the general public exposure, occupational exposure and the equipment and usage parameters.

There are countless man-made sources of electromagnetic radiation (EMR), such as mobile phones [5,6,7,8], radios and high-frequency (HF) welding machines. Aboard marine ships, there are numerous sources of EMR [9,10,11]; these include transmitters, such as radar and satellite for shipboard communication. Because of the extreme growth of the marine transportation industry, the number of marine vessel transmitters has greatly increased. Measuring the electric fields from marine ship transmitters is normally a complex and time-consuming process. Hence, there is a potential applicability of this work to a variety of marine vessels, which is significant, and could be quite useful as a reference for many professionals without direct access to the necessary equipment. Very high-frequency (VHF) radios [9], working in a frequency range from 156 MHz–176 MHz [10], and radars [11] seem to be more common in marine vessels that have been investigated in previous studies. The electromagnetic exposures aboard vessels could be affected by: (i) on-board radiation sources; (ii) conductive objects; and (iii) neighbouring radiation sources, such as other vessels and coastal transmitters. Electromagnetic fields can be reduced by properly designing the systems and equipment, such as choosing the correct material for vessels and finding the correct positioning of equipment on-board [12].

The aim of this study was to measure and analyse the radio frequency radiation emitted from transmitters aboard a marine vessel, focused on areas normally occupied by crew members and passengers; and to observe whether the measured electric field (EF) levels comply with the ICNIRP and the ARPANSA limits. In this analysis, we used transmitters: VHF band fixed (156–157.42 MHz), MF/HF transceiver (1605 KHz–30 MHz), satellite communication (Sat-Com C) (6006) (1,626.5–1,646.5 MHz), shipborne Automatic Identification System (AIS) navigation (156–163 MHz), radar X-band (9410 MHz) and radar S-band (3,050 MHz).

2. MATERIALS AND METHODS

The levels of electric field were recorded in all areas where crew members and passengers gather in all three levels of the ship: (i) the upper deck, Level 1 (the area where the passengers spend some of their time); (ii) the bridge deck, Level 2 (central to where the bridge operation stations are located and the crew members' working position); and (iii) the bridge roof, Level 3 (where all of the transmitters and antennas were located), as in Figure 1. Please note that the bridge roof is not accessible to the general public (e.g., passengers). The shipboard radio transmitter and the antenna systems that we used in this study are described in Table 1. The position numbers or measuring locations in Table 2 correspond to the numbers in Figure 1.

All electric field measurements were recorded under similar conditions with (i) a NARDA NBM 520 with probe (EF 1891), frequency range 3 MHz–18 GHz and sensitivity 0.8–1,000 V/m, and (ii) a NARDA NBM 550 with probe (EF 0391), frequency range 100 kHz–3 GHz, sensitivity 0.2 V/m for the electric field and a detection range 0.2–320 V/m (Narda Safety Test Solutions, Germany), to measure the electric field levels in this study. Both broadband and narrowband (frequency selective) instruments can be used for measuring RF fields. In our study, the NARDA meters provide the broadband measurements, and the electric field probes provide isotropic (non-directional) measurements.

The data acquisition time of the NARDA meters is 2 s–5 min. Each position undertook two types of measurements, "spot measurements" and "continuous measurements". Each set of electric field measurements was taken for 6 minutes at each position until a steady-state value was obtained; average values were then recorded. All probes were oriented vertically during the data collection, and the auto-zeroing function was used to exclude the effects of temperature on the results. Since the field distribution is not homogeneous, the spatial averaging function was used in both meters. The transmission directions of directional antennas did not change during the data collection period.

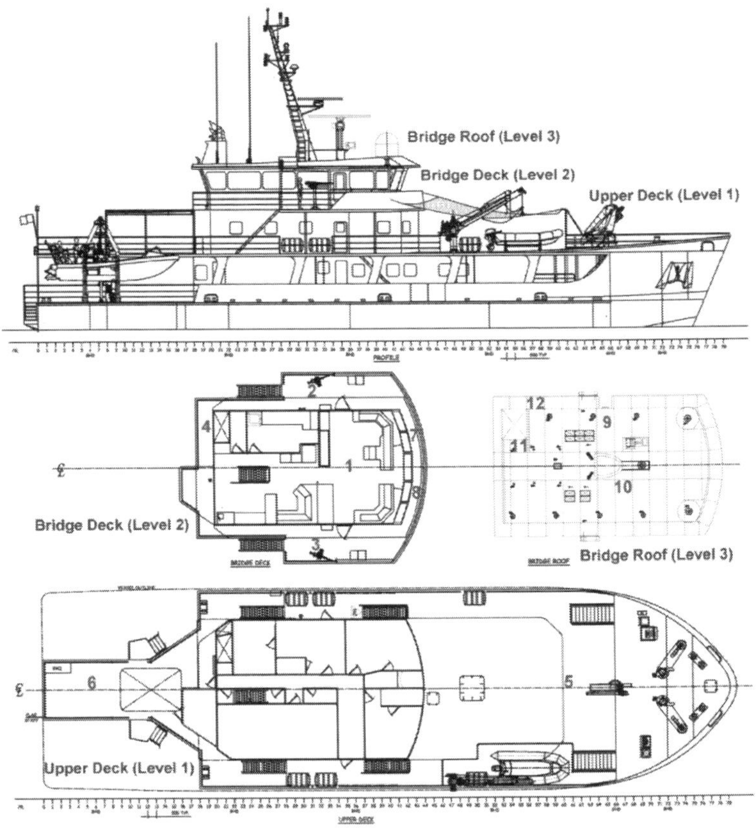

Figure 1. Measuring locations (No. 1–12) of the marine ship. The electric field (EF) measurements were recorded in all areas where people (crew members and passengers) gather in all 3 levels of the ship: (i) the upper deck (Level 1, the area where the passengers and crew members spend time); (ii) the bridge deck (Level 2, the crew members' working position); and (iii) the bridge roof (Level 3, all the transmitters and antennas were mounted). Please note that the bridge roof is not accessible to the general public (e.g., passengers).

Table 1. Details of shipboard radio transmitter and antenna systems measured. VHF, very high frequency; MF, medium frequency; HF, high frequency.

Radio Transmitter	Manufacturer/Model	Frequencies	Antennas	Radiation Hazard Distance from Transmitter	License
VHF Fixed	ICOM IC-M505	156.0–157.425 MHz	GFE	1 m	Under Ships Marine Radio License
VHF Fixed	Sailor 6006	156.0–157.425 MHz	GFE	1 m	Under Ships Marine Radio License
			GMDSS Suite		
MF/HF Transceiver	Sailor	1605 KHz–30 MHz	ATU6381 Coupler with ANT 800W SSB-HF 8.5MT. (Rx/Tx) ANT 800W SSB-HF 7.0MT. (Tx)	1.8 m	Under Ships Marine Radio License
Satellite Communication (Sat-Com C)	Sailor 6006	1626.5–1646.5 MHz	TT-3027C	0.3 m	Under Ships Marine Radio License
			Navigation System		
Automatic Identification System (AIS) Navigation	SAAB/R5 AIS	156–163 MHz	Celmar	0.5 m	Under Ships Marine Radio License
Radar X-band (25 kW X-band transceiver)	Sperry Marine/VMFT ECAT2 253/8/MK/VM2	9,410 MHz	Included, 2.5 m	100 W/m² for 0.75 m 10 W/m² for 7.5 m	Under Ships Marine Radio License
Radar S-band (30 kW S-band transceiver)	Sperry Marine/VMFT ECAT2 253/8/MK/VM2	3,050 MHz	Included, 3.5 m	100 W/m² for 0.55 m 10 W/m² for 5.5 m	Under Ships Marine Radio License

Table 2. Measuring locations in all 3 levels of the ship.

Location	Reason for the choice of location
	Bridge Deck, Level 2
No. 1. Central to the bridge operation stations	The crew members' working position
No. 2. The port wind station	Between the gun and the wind station control (on the left)
No. 3. The port wind station	Between the gun and the wind station control (on the right)
No. 4. The observation deck	Behind the bridge deck
No. 7. The bridge deck and bridge windows	Opposite the crew members' working position (on the left)
No. 8. The bridge deck and bridge windows	Opposite the crew members' working position (on the right)
	Upper Deck, Level 1
No. 5. Passenger area	Frequently used by passengers
No. 6. The gymnasium	The crew members spend some of their time here
	Bridge Roof, Level 3
No. 9. The top of the access ladder onto the bridge roof	Bridge staff spend their time here
No. 10. Beside the ship mast, which is at the bottom of the access ladder up the ship mast	Some antennas and equipment were mounted on the mast
No. 11. Beside the panels and close to the ventilation flats	This is in the shadow of the ship mast
No. 12. Close to the ventilation flats	Outside the shadow sector of the ship mast

2.1. Spot and Continuous Measurements

Electric field measurements were then taken while the ship was not moving (in the "standing" position), with the engine turned off. Around each of the locations, the meters were moved slowly along the whole area to identify the typical and highest electric field (EF) levels, as well as the location of the high electric fields. Because the likelihood of peaks and nulls in the fields was very high in the confines of the metal ship, the spot measurements were conducted to identify sources of EF. Then, the continuous measurements were taken from the identified locations. Please note that the EF measurements were taken during the weekend to minimise the impact of external disturbances.

Firstly, all of the transmitters and antennas that were installed on the bridge roof were turned off to measure the background electric field levels. Secondly, spot and continuous measurements were recorded in a 40 m-long ship (in total, 528 electric field measurements) [13]. The ship was positioned in a port and about 150 m from the nearest vessel. Measurements were carried out at 12 different locations (Table 2) and under 11 different scenarios (Table 3) in all 3 levels of the ship (upper deck, bridge deck and bridge roof) in locations where crew members and passengers spend their time. The appropriate source of antenna tuning was completed by the radio operator to obtain the maximum level of radiated power. Due to the port emission restriction, these transmitters were switched on only when each location was measured. No other people were on the ship during the measurement period. Repeated measurements were taken at each location to ensure the validity of the data.

2.2. Data Collection Protocol: Various Scenarios of RF Transmitters on Marine Vessels and Reasons

Under a normal operation condition, the ship will have the two radars (X-band and S-band), the AIS and the Sat Comm C; all transmitting continuously. The VHF and the MF/HF radios operate intermittently, in addition to the other navigation equipment transmitting continuously. By considering all of these situations, we developed a data collection protocol and performed various scenarios to observe and measure the radiation emissions from all of the transmitters. The protocol and reasons for the various scenarios are described in Table 3.

Table 3. Data collection protocol: the various scenarios of marine ship transmitters for measuring RF-EF levels.

Scenario No.	Scenario Details
A	Measured with the VHF fixed (IC-M505) switched ON and all other transmitting antennas switched OFF
B	Measured with the VHF fixed (Sailor 6006) switched ON and all other transmitting antennas switched OFF
C	Measured with the MF/HF transceiver (TU 6360 TX/Tx and CU6301 control unit) switched ON and all other transmitting antennas switched OFF
D	Measured with the satellite communication, Sat-Com C (6006), switched ON and all other transmitting antennas switched OFF
E	Measured with the AIS navigation (R5 AIS) switched ON and all other transmitting antennas switched OFF
F	Measured with the radar X-band (VMFT ECAT2 253/8/MK/VM2) switched ON and all other transmitting antennas switched OFF
G	Measured with the radar S-band (VMFT ECAT2 253/8/MK/VM2) switched ON and all other transmitting antennas switched OFF
H	Measured with the Group Z (radar X-band, radar S-band, satellite communication (Sat-Com C) and AIS navigation) switched ON and all other transmitting antennas switched OFF (under the normal operation condition, the ship will have two radars, the AIS and Sat Comm C; all transmitting continuously; besides, Sat Comm C does not continuously transmit; however, it operates all of the time; the VHF and MF/HF radios operate intermittently, in addition to the other navigation equipment transmitting continuously)
I	Measured with the Group Z + VHF fixed (IC-M505) switched ON
J	Measured with the Group Z + VHF fixed (Sailor 6006) switched ON
K	Measured with the Group Z + MF/HF transceiver (TU 6360 TX/Tx and CU6301 control unit) switched ON

2.3. Data Analysis Using Occupational and General Public Exposure Levels

First, we analysed the environment or background EF levels, where all of the transmitters and antennas that were installed on the bridge roof were turned off at each measurement point. As the EF levels were not measured on the same day for all transmitters, the background fields were measured on each day. On average, those values were observed ranging from 0.000 to 0.003 V/m. The background electric field values at different locations were subtracted from all of the other measured data, and negative numbers were rounded to zero.

The probability distribution of the EF levels was observed in order to test the normality of the distribution. The sample sizes used were: for the bridge roof N = 176, the bridge deck N = 264 and the upper deck N = 88. For each set of measurements (one location at one data collection protocol), we calculated the mean and the standard deviation of each dataset. The geometric mean is more applicable than the arithmetic mean (AM) for explaining the relative progress of the EF level values. Therefore, we computed the geometric mean (GM) of each dataset to estimate the population. Nevertheless, the percentage of measurements equal to or greater than 14 V/m (50% of the ICNIRP exposure limit) and 28 V/m (the ICNIRP exposure limit) were calculated.

The interquartile range (IQR) is a vital estimate of the spread of the data, as changes in the upper and lower 25% of the data do not affect it. As an estimate, the IQR is more representative than the standard deviation of the spread of the data, if there are outliers. When the data are from a normal distribution, the IQR is less efficient than the standard deviation; however, we found that our data did not have a normal distribution. Therefore, we analysed the interquartile range of the datasets to perceive how data spread over the 25th (first quartile) and 75th (third quartile) percentiles.

The EF measurements were analysed and compared in all 3 levels of the ship (upper deck, bridge deck and bridge roof) and the operating conditions of the vessels' transmitters and antennas. For comparison, we used box-and-whiskers plots to show these values. Then, we compared the exposure values with the general public exposure levels of the ICNIRP and the ARPANSA limits in order to quantify the effects. All analyses were carried out using MATLAB (MathWorks Inc., Natick, MA, USA) R2014b on a computer with an Intel Core Intel Core i7 CPU.

Table 4. The occupational and general public exposure levels of the ICNIRP guidelines and the ARPANSA standards.

Frequency	General Public (V/m)	Occupational (V/m)
	ICNIRP Guidelines	
100 kHz–3 GHz	28.0–87.0	61.0–610.0
3 MHz–18 GHz	28.0–61.0	61.0–203.03
	ARPANSA Standards	
100 kHz–3 GHz	27.4–86.8	61.4–614.0
3 MHz–18 GHz	27.4–61.4	61.4–204.6

The regulation and recommendations about the health and environmental effects on all aspects of non-ionizing radiation exposure was given by the independent scientific organization, the International Commission on Non-Ionizing Radiation Protection (ICNIRP) [1]. In Australia, the occupational and general public exposure levels are mandated by the Australian Radiation Protection and Nuclear Safety Agency (ARPANSA) standards: "Radiation Protection Standard for Maximum Exposure Levels to Radio frequency Fields–3 kHz to 300 GHz", 2002 (ARPANSA Radiation Protection Standard No. 3 (RPS3)) [4]. These documents provide reference levels, hence providing verification of compliance with the basic restrictions of the standards. This requires measurements of the highest RF field levels emitted under normal operating conditions and the maximum expected duty factor in areas accessible to workers or the general public [4]. The time-averaged electric field levels for both public and occupational levels are given in Table 4. In the next section, the results of the measurements are described and compared to the reference levels of the ICNIRP guidelines and the ARPANSA standards.

3. RESULTS AND DISCUSSION

The measured electric fields were then assessed for compliance with the occupational and general public reference levels of the ICNIRP guidelines [1] and the ARPANSA standards [4]. Overall, the average electric field levels for various marine vessel's transmitters were in the range of 0.55–7.99 V/m (AM) and 0.47–3.70 V/m (GM). These fields were below the ICNIRP guidelines, 28 V/m [1], and the ARPANSA limits, 27.4 V/m [4], for maximum general public exposure.

Figure 2, Figure 3 and Figure 4 show the levels of the electric fields from the marine ship transmitters using box plots with a median and interquartile range. The T-bars (whiskers) were extended to 1.5-times the height of the box (IQR). A "+" symbol depicts a value more than three IQRs from the end of the box (the extreme outlier). A few points with high electric field strengths that exceeded the ICNIRP and the ARPANSA limits for the general public were observed on the bridge roof, as in Figure 2; nevertheless, the occupational limits were respected everywhere. Thus, this complies with the occupational and general public reference levels of the ICNIRP guidelines and the ARPANSA standards.

Table 5 presents the descriptive statistics for the electric field measurements of the different levels of the ship. The mean EF levels were significantly (6.321 V/m) higher on the bridge roof compared to the other levels of the ship. This is obvious as all of the transmitters were mounted at Location Nos. 9–12, and this corresponds to Figure 2. The 75th percentile of the EF measurements was also high (9.377 V/m) on the bridge roof level, and the percentage of measurements above 27.4 V/m were slightly high (3.41%) at this level. The electric field levels were significantly lower in the upper deck level compared to the bridge deck level. Despite that, the general public has no access to the bridge roof at Location Nos. 9–12; therefore, the bridge roof data were assessed for compliance with the occupational reference levels (61 V/m), while upper deck and bridge deck data were assessed by the general public reference levels (28 V/m). All of the measured electric field data were compared to the lowest limit of the respective reference levels, as indicated in Table 4.

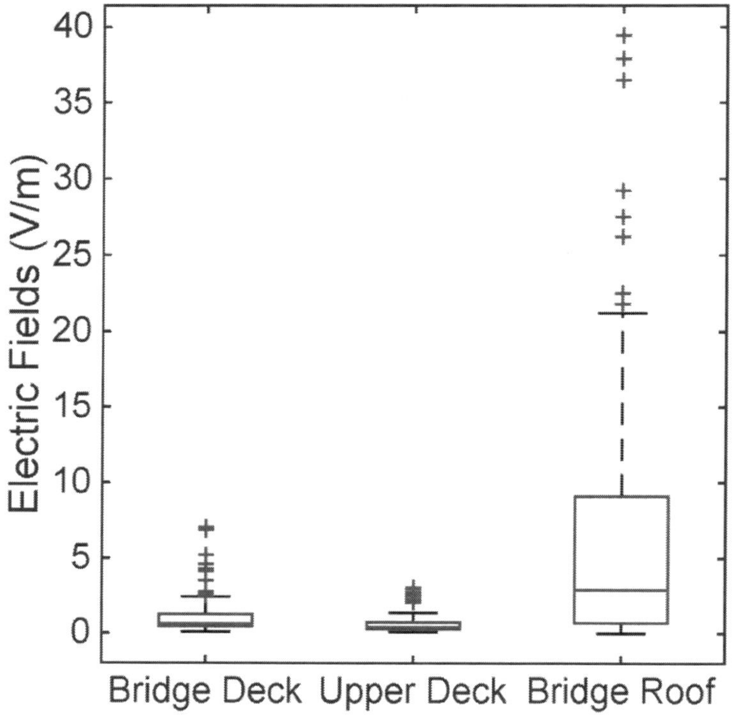

Figure 2. Electric field level measurement of the bridge deck, upper deck and bridge roof using box plots with a median and interquartile range. The T-bars (whiskers) were extended to 1.5-times the height of the box (inter quartile range, IQR). A "+" symbol depicts a value more than three IQRs from the end of the box (the extreme outlier). The ICNIRP and the Australian Radiation Protection and Nuclear Safety Agency (ARPANSA) limits for the general public were exceeded on the bridge roof; nevertheless, the occupational limits were respected everywhere.

Table 6 shows the electric field levels in different locations on the ship. The mean EF levels were significantly higher (7.992 V/m) than in Location No. 12 on the bridge roof. The 75th percentile of the EF measurements was also high (12.352 V/m) in Location No. 12, and the percentage of measurements above 27.4 V/m was also slightly high (4.54%). The electric field levels were also significantly higher in Location Nos. 9–12 compared to the others. This is obvious, as all of the transmitters were mounted on the bridge roof. Figure 3 shows the electric field variation based on the locations that are described in Table 2 and the values obtained from Table 6.

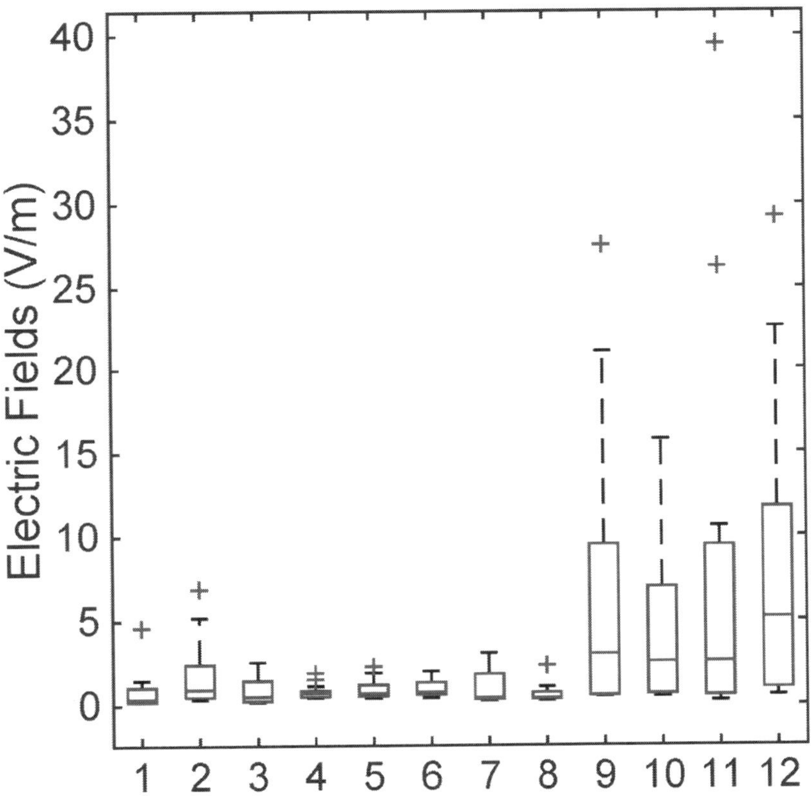

Figure 3. Electric field level measurement by different locations, as in Table 2 using box plots with a median and interquartile range. The T-bars (whiskers) were extended to 1.5-times the height of the box (IQR). A "+" symbol depicts a value more than three IQRs from the end of the box (the extreme outlier). The ICNIRP and the ARPANSA limits for the general public were exceeded on the bridge roof; nevertheless, the occupational limits were respected everywhere.

Table 7 shows the electric field levels of various scenarios that are described in Table 3. The mean EF levels were significantly (6.734 V/m) high in Scenario I, when the radar X-band, radar S-band, satellite communication (Sat-Com C), AIS navigation and VHF fixed (IC-M505) were switched on. The 75th percentile of the EF measurements was also high (12.352 V/m) in Scenario J, where the radar X-band, radar S-band, satellite communication (Sat-Com C), AIS navigation and VHF fixed (Sailor 6006) were switched on. The percentage of measurements above 27.4 V/m was slightly higher (8.33%) than in Scenario I. The electric field levels were also significantly higher in Scenarios H–K compared to the others. Similarly, Figure 4 shows that the electric field level variation is based on the various scenarios that are described in Table 3 and the values obtained from Table 7.

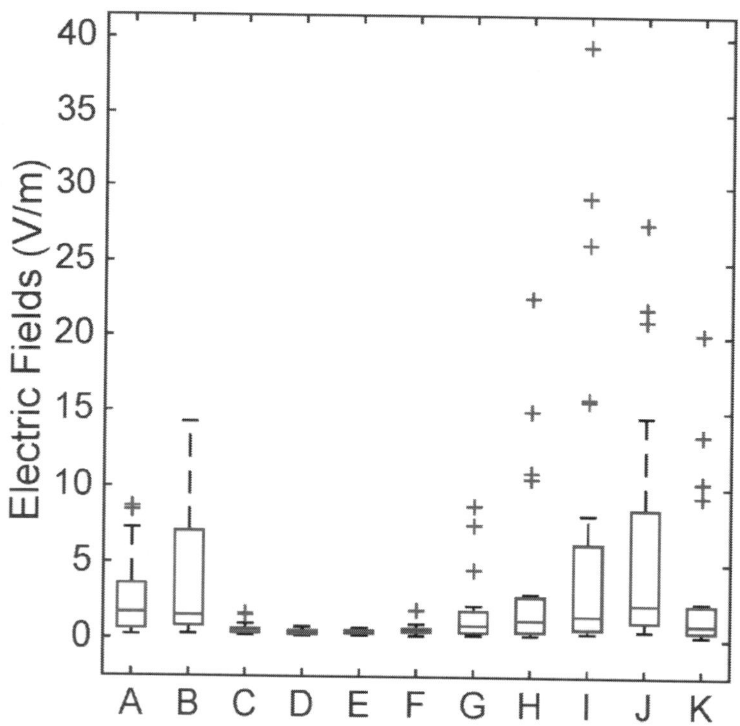

Figure 4. Electric field level measurement by various scenarios as in Table 3 using box plots with a median and interquartile range. The T-bars (whiskers) were extended to 1.5-times the height of the box (IQR). A "+" symbol depicts a value more than three IQRs from the end of the box (the extreme outlier). The ICNIRP and the ARPANSA limits for the general public were exceeded on the bridge roof; nevertheless, the occupational limits were respected everywhere.

Repeated measurements resulted in average electric field levels that differed from each other by about ±11%. Nonetheless, we observed that the effect for crew members and passengers from marine vessel transmitters is minimised from the measurements that were carried out under 11 different scenarios and at 12 different locations on the vessel. Our results demonstrated that the electric field levels measured complied with the international (ICNIRP) [1] and the Australian (ARPANSA) standards [4] and were also compatible with the measurements recorded in a previous study [9]. Nevertheless, for the ICNIRP limit, the acute effects are determined by the intensity of the radiation, and cumulative effects were not presumed, as also indicated in our previous research [14]. For low frequencies, a limit of the magnetic field could be exceeded even if the electric field limit is low. In contrast to this, at high frequencies (>100 KHz), magnetic and electric fields are decoupled when the near-field is considered.

Table 5. The descriptive statistics for electric field levels (V/m) emitted from marine transmitters from each ship level.

Location	N	Minimum	Maximum	Mean (SD)	Geometric Mean (GSD)	Median	25th–75th Percentile	% >14 V/m	% >27.4 V/m	Compliance
Overall	528	0.001	39.460	2.631 (4.925)	1.027 (3.744)	0.755	0.440–2.21	3.97	0.94	Yes
Bridge Roof	176	0.001	39.460	6.001 (7.348)	2.537 (4.959)	2.945	0.702–9.105	11.93	2.84	Yes
Bridge Deck	264	0.070	7.040	1.029 (1.037)	0.723 (2.310)	0.630	0.442–1.267	0.00	0.00	Yes
Upper Deck	88	0.070	3.090	0.694 (0.729)	0.480 (2.237)	0.400	0.302–0.747	0.00	0.00	Yes

Table 6. The descriptive statistics for electric field levels (V/m) emitted from the marine transmitters from each location of the ship.

Location No	Minimum	Maximum	Mean (SD)	Geometric Mean (GSD)	Median	25th–75th Percentile	% >14 V/m	% >27.4 V/m	Compliance
1	0.210	4.610	0.753 (0.959)	0.491 (2.311)	0.395	0.230–1.065	0.00	0.00	Yes
2	0.360	6.870	1.777 (1.755)	1.157 (2.511)	0.935	0.497–2.513	0.00	0.00	Yes
3	0.180	2.580	0.913 (0.831)	0.609 (2.479)	0.510	0.255–1.467	0.00	0.00	Yes
4	0.440	1.910	0.782 (0.371)	0.718 (1.479)	0.700	0.497–0.890	0.00	0.00	Yes
5	0.400	2.300	0.962 (0.590)	0.826 (1.699)	0.695	0.540–1.235	0.00	0.00	Yes
6	0.410	2.020	0.971 (0.488)	0.868 (1.594)	0.735	0.587–1.377	0.00	0.00	Yes
7	0.250	3.090	0.985 (0.950)	0.657 (2.372)	0.415	0.325–1.877	0.00	0.00	Yes
8	0.250	2.330	0.555 (0.425)	0.471 (1.670)	0.380	0.327–0.737	0.00	0.00	Yes
9	0.470	27.520	6.328 (7.645)	2.766 (3.982)	3.010	0.540–10.447	18.18	4.54	Yes *
10	0.490	15.790	4.440 (4.748)	2.332 (3.339)	2.545	0.627–7.085	9.09	0.00	Yes
11	0.230	39.460	6.523 (9.617)	2.158 (5.037)	2.565	0.505–9.622	0.09	4.54	Yes *
12	0.540	29.230	7.992 (8.656)	3.709 (3.914)	5.185	0.925–12.352	22.73	4.54	Yes *

* The limits for the general public (e.g., passengers) were exceeded on the bridge roof, and the occupational limits were respected everywhere.

Table 7. The descriptive statistics for electric field levels (V/m) emitted from the marine transmitters from each scenario.

Scenario No	Transmitter	Minimum	Maximum	Mean (SD)	Geometric Mean (GSD)	Median	25th–75th Percentile	% >14 V/m	% >27.4 V/m	Complianc
A	VHF band fixed (IC-M505)	0.240	8.670	2.629 (2.593)	1.582 (2.916)	1.700	0.603–7.107	0.00	0.00	Yes
B	VHF band fixed (Sailor 6006)	0.290	14.250	3.855 (4.117)	2.066 (3.250)	1.525	0.820–7.107	4.17	0.00	Yes
C	MF/HF transceiver	0.210	3.420	0.585 (0.361)	0.505 (1.691)	0.540	0.342–0.637	0.00	0.00	Yes
D	Satellite communication	0.000	0.790	0.445 (0.162)	0.415 (1.479)	0.490	0.295–0.540	0.00	0.00	Yes
E	AIS navigation	0.070	0.720	0.456 (0.144)	0.432 (1.409)	0.490	0.340–0.570	0.00	0.00	Yes
F	Radar X-band	0.180	4.250	0.595 (0.330)	0.531 (1.590)	0.540	0.457–0.700	0.00	0.00	Yes
G	Radar S-band	0.070	12.450	1.958 (2.490)	1.081 (2.823)	0.895	0.422–1.915	0.00	0.00	Yes
H	Group Z	0.070	22.540	3.429 (5.599)	1.416 (3.610)	1.245	0.462–2.840	8.33	0.00	Yes
I	Group Z + VHF fixed (IC-M505)	0.160	39.460	6.734 (10.706)	2.345 (4.094)	1.55	0.640–7.217	20.83	8.33	Yes *
J	Group Z + VHF fixed (Sailor 6006)	0.560	27.520	6.227 (7.690)	3.170 (3.170)	2.300	1.110–9.390	16.66	4.16	Yes *
K	Group Z + MF/HF transceiver	0.230	20.160	3.144 (5.031)	1.332 (3.386)	0.970	0.512–2.302	4.16	0.00	Yes

* The limits for the general public (e.g., passengers) were exceeded on the bridge roof, and the occupational limits were respected everywhere.

Larger surveys are desired to verify our findings and to provide reliable knowledge on radio hazard safety assessment for marine ship transmitters and the electromagnetic compatibility (EMC). As indicated in [15], the sensitivity of the location and standardised and well-described data collection settings are significantly important and should be considered in the comparison of different studies. To reduce electromagnetic fields, marine vessel designers may use low permeable materials, such as glass reinforced plastic (GRP), wood, aluminium hulls in the vessel and equipment construction [16]. The magnetic fields reduce rapidly with distance; hence, equipment containing inevitable magnetic sources, such as electric motors and transformers, should be located as far away as possible from areas normally occupied by crew members and passengers.

4. CONCLUSIONS

In this investigation, we performed measurements and analysed radio frequency radiation emitted by the transmitters aboard a marine vessel, focusing on areas normally occupied by crew members and passengers. In total, 528 electric field measurements were taken. Additionally, we developed a new data collection protocol and performed various scenarios to accurately measure the radiation from all transmitters. Under the normal operating conditions, there were a few marine ship transmitters and antennas transmitting continuously, and other radios operate intermittently. By considering this, for the first time, we report measuring the electric field from each transmitter condition, which is insignificant, and this must be carefully taken into account for future studies. Our results show that the electric field levels were highest on the bridge roof and the lowest in the upper deck, and the measured values were within a range of 0.001–39.46 V/m. The limits for the general public were exceeded on the bridge roof; nonetheless, the occupational limits were respected everywhere. Hence, this complies with the occupational and general public reference levels of the ICNIRP guidelines and the ARPANSA standards. Some further conclusions that can be drawn from this paper are: (i) electric field levels were high with the VHF fixed (Sailor 6006) transmitter; and (ii) high frequency electric field levels that are radiated from the vessels' transmitters on the bridge roof will not have much impact for crew members and passengers. Nevertheless, this study should be useful as a reference for many researchers and industry professionals without direct access to the necessary equipment. Further research is desired to determine the electric field levels for a larger amount of ships using the proposed protocol in this paper. Such research would provide a basis for establishing safety distances and support the development of guidelines by suitable authorities.

ACKNOWLEDGEMENTS

The author would like to thank Andrew Ross for the excellent technical assistance during the data collection and Lyn Mclean from Electromagnetic Radiation (EMR) Australia Private Limited (PL).

CONFLICTS OF INTEREST

The authors declare no conflict of interest.

REFERENCES

1. ICNIRP (international commission on non-ionizing radiation protection). Guidelines for limiting exposure to time-varying electric, magnetic and electromagnetic fields (up to 300 GHz). Health Phys. **1998**, 74, 494–522.

2. IEEE. C95.1-2005—IEEE Standard for Safety Levels with Respect to Human Exposure to Radio Frequency Electromagnetic Fields, 3 KHz to 300 GHz; Institute of Electrical and Electronics Engineers: Piscataway, NJ, USA, 2005.

3. CENELEC EN 50413. Basic Standard on Measurement and Calculation Procedures for Human Exposure to Electric, Magnetic and Electromagnetic Fields (0 Hz–300 GHz); Techicnal report; CENELEC: Brussels, Belguim, 2008.

4. Australian Radiation Protection and Nuclear Safety Agency. Maximum Exposure Levels to Radiofrequency Fields–3 kHz to 300 GHz; Australian Radiation Protection and Nuclear Safety Agency (ARPANSA): Melbourne, Australia, 2002.

5. Hirose, H.; Suhara, T.; Kaji, N.; Sakuma, N.; Sekijima, M.; Nojima, T.; Miyakoshi, J. Mobile phone base station radiation does not affect neoplastic transformation in BALB/3T3 cells. Bioelectromagnetics **2008**, 29, 55–64.

6. Grafström, G.; Nittby, H.; Brun, A.; Malmgren, L.; Persson, B. R.; Salford, L.G.; Eberhardt, J.L. Histopathological examinations of rat brains after long-term exposure to GSM-900 mobile phone radiation. Brain Res. Bull. **2008**, 77, 257–263.

7. Halgamuge, M.N.; Yak, S.K.; Eberhardt, J.L. Reduced growth of soybean seedlings after exposure to weak microwave radiation from GSM 900 mobile phone and base station. Bioelectromagnetics **2015**, 36, 87–95.

8. Halgamuge, M.N. Radio Frequency Electromagnetic Fields from Mobile Phone. In Electromagnetics Engineering Handbook: Analysis and Design of Electrical and Electronic Devices and Systems,

Handbook of Applied Electromagnetics; WIT Press: Ashurst, UK, 2013; Volume 1, pp. 383–408.

9. Senic, D.; Sarolic, A. Simulation of a shipboard VHF antenna radiation pattern using a complete sailboat model. In Proceedings of the IEEE Intternational Conference on Software, Telecommunications and Computer Networks, Hvar, Croatia, 24–26 September 2009; pp. 65–69.

10. Senic, D.; Sarolic, A. Simulation of slanted shipboard VHF antenna radiation pattern. In Proceedings of the 52nd International Symposium ELMAR-2010, Zadar, Croatia, 15–17 September 2010; pp. 293–296.

11. Moen, B.; Møllerløkken, O.J.; Bull, N.; Oftedal, G.; Mild, K.H. Accidental exposure to electromagnetic fields from the radar of a naval ship: a descriptive study. Int. Marit. Health **2013**, 64, 177–182.

12. Hasril, N.; Mahdi, C.; Muhammad, M.; Hassanuddin, N.; Yusoff, N.; Yati, M.; Nor, I. Management of naval vessels' electromagnetic signatures: a review of sources and counter measures. Defence S & T Tech. Bull. **2013**, 6, 93–110.

13. Halgamuge, M.N. Analysis of shipboard high frequency electromagnetic fields. Radiat. Prot. Australas. **2014**, 31, 13–24.

14. Halgamuge, M.N. Pineal melatonin level disruption in humans due to electromagnetic fields and ICNIRP limits. Radiat. Prot. Dosimetry **2013**, 154, 405–416.

15. Hareuveny, R.; Sudan, M.; Halgamuge, M.N.; Yaffe, Y.; Tzabari, Y.; Namir, D.; Kheifets, L. Characterization of extremely low frequency magnetic fields from diesel, gasoline and hybrid cars under controlled conditions. Int. J. Environ. Res. Public Health IJERPH **2015**, 12, 1651–1666.

16. Diaz, E.S.; Tims, R. A complete underwater electric and magnetic signature scenario using computational modelling. In Proceedings of the Marine Electrommagnetics Conference (MARELEC), Amsterdeam, Netherlands, 19–21 April 2006.

CHAPTER 11

Naval Surface Ship In-service Information Exploitation

Gary Ford[a]*, Chris McMahon[a], Chris Rowley[b]

[a] University of Bristol, Bristol BS8 1UB, England

[b] Babcock International, Keynsham BS31 2YH, England

ABSTRACT

The Royal Navy operates a fleet of complex modern warships and submarines each comprising a system of systems often in harsh and potentially volatile environments. The maintenance of surface vessels is primarily undertaken by Babcock and BAe Systems in an alliance with the Ministry of Defence. The Ministry of Defence system engineering lifecycle is known as CADMID, this details the six phases of a projects' lifecycle from Concept through to Disposal. The "In-Service" phase of a naval vessel will typically constitute 70% of the artefact's through-life cost. During the "In-Service" phase the number and involvement of stakeholders will vary as the vessel cycles through Tasking, Upkeep and Regeneration. The paper considers the key stakeholders and their participation in each cyclical mode. Information to be exploited will be subject to two discrete drivers, firstly the information available for exploitation as a consequence the vessels' current cyclical mode, secondly, the characteristics of the information source..

KEYWORDS

Naval; In-Service; Maintenance; Information; Exploitation;

1. INTRODUCTION

A modern warship / submarine will contain in excess of 100 integrated systems which are linked structurally, mechanically, electrically, hydraulically, pneumatically and electronically [1]; thus a warship / submarine may be viewed as a system of systems. Maintaining and enhancing the capability of modern warships and submarines is a strategic function undertaken by organisations working in partnership with the Royal Navy (RN) capable of providing "ultra-reliable engineering excellence". The maintenance of warships and submarines often necessitates the exploitation of vast quantities of information, Kane and Alavi define "exploitation" as "incremental learning focused on diffusion, refinement, and reuse of existing knowledge" [2], indeed, often what is learnt is an incremental enhancement of knowledge already gained. However, caution is fundamental when endeavouring to exploit information since it may exhibit some or all of the following [3]:

- Fuzziness: lack of detail
- Incompleteness: that which we do not know or leave out
- Randomness: lack of pattern

The approach of this paper is as follows. After outlining the background of the marine defence environment within the United Kingdom (UK) it explores the systems engineering methodologies applied to the domain followed by a review of "objective" and "subjective" information sources. The "In- Service" stakeholders are reviewed together with their association with the cyclical operational modes of surface ships and this is followed by exploration of the information available for exploitation and the associated characteristics of this information. The paper concludes with a discussion, further research and conclusion.

The paper presents a position statement, taken from the experience of naval and Babcock personnel within the surface ship domain; the purpose is to identify stakeholders, information sources and associated characteristics currently available and exploited.

2. BACKGROUND

The RN fleet comprises approximately 95 vessels [4] varying in size and complexity, however, "taking the number of frigates and destroyers as an indicator, the RN is one quarter of the size it was in 1980" [5]. None-the-less, operational commitments remain high, with on-going operations in the Arabian Gulf, Indian Ocean and the South Atlantic in addition to providing the UK's continuous nuclear deterrence. The surface fleet is supported by a Surface Ship Support Alliance (SSSA), formed between Babcock, BAe Systems (BAe) and the Ministry of Defence (MOD) creating a Class Output Management (COM) organisation to maintain surface ships, based upon the premise of "contracting for availability" [6].

Naval vessels, i.e. surface ships and submarines, are data rich environments containing and creating large volumes of electronic and paper records, e.g. Operational Defects (OpDef), Marine Engineering Defect logs, etc. Each data source is often designed to satisfy a singular requirement within a singular cyclical mode; the consequence being the risk of data duplication and potentially a failure to record data that may be pertinent to exploiting the vessels' / system material state.

In addition to the numerous data sources local management will often vary, e.g. some vessels may record all defects in the maintenance management system, i.e. Unit Maintenance Management System (UMMS), whereas other vessels may only transpose into UMMS the most persistent defects as logged in the engineering logs

3. LITERATURE REVIEW

The International Council on Systems Engineering (INCOSE) defines a system as, "an integrated set of elements, subsystems, or assemblies that accomplish a defined objective" [7]. Naval vessels are complex and may be viewed as a system of systems, i.e. interrelated systems with emergent properties.

A variety of systems engineering standards exist (Table 1); the variation in definition and application of each standard reflects "the nature, purpose, use and prevailing circumstances of the system" [8].

Table 1. Systems Engineering Standards.

Standard (Year)	Description
ANSI/EIA-632 (1999)	Processes for Engineering a System
IEEE-1220 (1998)	Application and Management of the Systems Engineering Process
ISO 15288 (2008)	Systems and software engineering - System life cycle processes
CMMI® (2002)	Capability Maturity Model® IntegrationSM (CMMI®)
MIL-STD-499C (2005)	Systems Engineering

The acquisition and support of RN vessels by the MOD is defined by the "Acquisitions Operating Framework" [9]. The development and acquisition of "equipment capability" follows CADMID, similar in principle to ISO 15288, i.e. "Systems Engineering - System Life Cycle Processes" [10]. The MOD framework comprises 6 discrete stages (Figure 1).

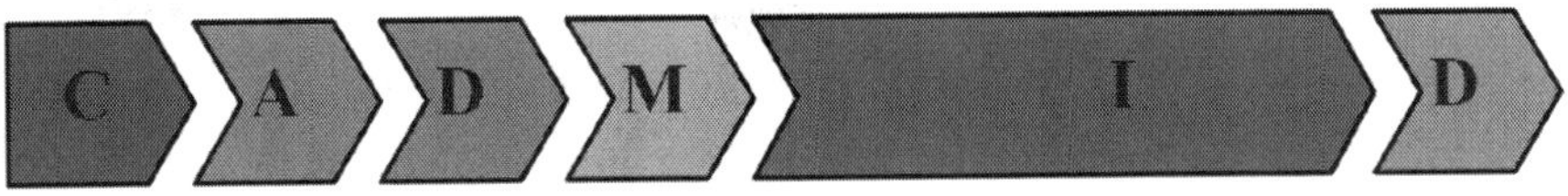

Figure 1. CADMID Lifecycle

The 6 stages are Concept, Assessment, Demonstration, Manufacture, In-Service and Disposal; thus a project / artefact should experience a linear progression through each stage of the lifecycle.

The In-Service phase of an RN vessel may be considerable, e.g. HMS Cornwall (Type 22 Frigate) was In-Service for more than 23 years. In order to sustain the capability of a naval vessel and extend its life, numerous Upkeep periods are undertaken which include capability upgrades, e.g. HMS Somerset experienced "enhancements to the ship's Seawolf missile system, installation of an advanced electronic communications system,..." [11].

Given the duration of the In-Service phase, the evolving operational threat and physical environment; timely effective maintenance together with prompt rectification of unexpected defects and system failure is essential to maintain the requisite level of capability and availability. Each Upkeep period will constitute a discrete CADMID lifecycle (Figure 2).

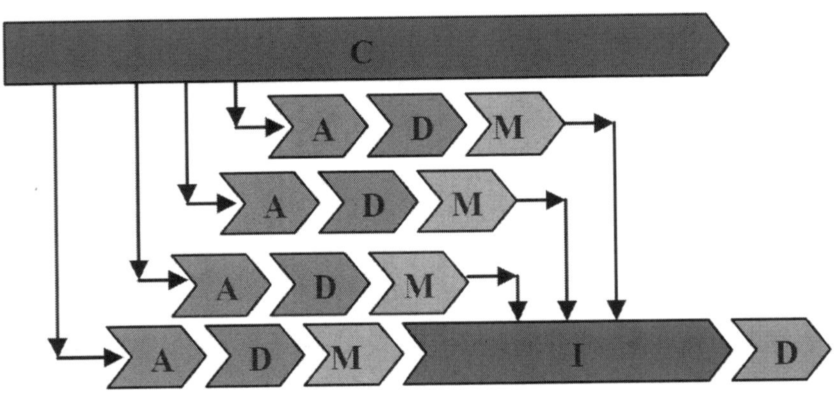

Figure 2. Capability Enhancement Lifecycle

The "Systems Engineering Handbook" [12] states a key benefit of systems engineering as "Ensuring stakeholder satisfaction on delivery". Indeed, failure to "effectively engage" with stakeholders is identified as a common cause for project failure [13]. Within the maritime defence environment engaging with and ensuring stakeholder satisfaction is the function of the "service provider" organisation, i.e. the COM, to the extent the customer (RN) is embedded within the structure.

The process model detailed in "A process view of maintenance and its stakeholders" [14] identifies 4 key stages,i.e. Maintenance Planning, Maintenance Execution, Functional Testing and Feedback. In addition to a definitive functional requirement, Soderholm asserts "Other inputs are maintenance documentation, such as maintenance objectives, strategies, and policies, which all should be based on stakeholder requirements" [14]. The "objectives, strategies, and policies" are essential to engender "customer" confidence with respect to the quality management process.

Stakeholders will often iterate the term "material state", however, an agreed and definitive definition is often more difficult to deduce. Giurgiutiu defines it as to "...characterize the material state by identifying critical features that show changes with respect to a reference state..." [15]. This would necessitate identifying the "critical features" and setting parameters beyond which degradation in performance or failure could be anticipated. In a complex system, the "critical features" are likely to be installation specific as a result of interconnectivity and emergent properties

A holistic assessment of the "material state" for a complex system is often not a binary value, e.g., available / unavailable; a complex system may be "available" but its efficiency / performance may be degraded or its ability to sustain further failure may be compromised. Information indicating the "material state" may be "objective" or "subjective", i.e. "existing in the mind or belonging to the thinking subject rather than to the object of thought" [3]. Information may be perceived as existing along a continuum of "subjective" / "objective"; indeed, "as we pass from assumption to assumption along the subjective-objective continuum, the nature of what constitutes adequate knowledge changes" [16]. In this regard the "objective" / "subjective" content of each information source will vary (Figure 3) and provide a particular point perspective of the "material state", e.g.

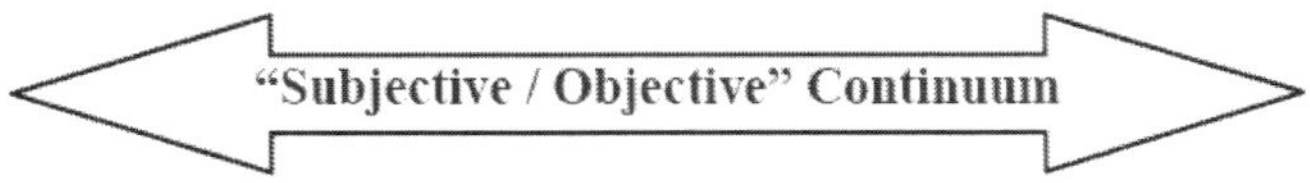

	Subjective	**Subjective / Objective**	**Objective**
Information sources	Operational Defect S2022 Defect report Metallurgy reports	Ships log	FLUBCON Material State Portfolio Test & trials reports Finance reports

Figure 3. "Subjective" / "Objective" continuum

Note: FLUBCON - Fuel and lubrication report

Information sources categorised as "subjective" will typically contain qualitative data, images, audio, … requiring interpretation, clarification. The information may contain a limited volume of quantitative data, e.g. engine running hours. "Objective" information sources will primarily be quantitative data, e.g. inlet / exhaust temperature. The source may contain a limited volume of qualitative data.

A balance (Figure 4) of "subjective" and "objective" information should provide the optimum perspective of a "material state".

The optimum balance of "subjective" and "objective" information will be dependent upon the information source(s) available, stakeholder requirement and the vessels mode, i.e. Tasking, Regeneration or Upkeep (see below).

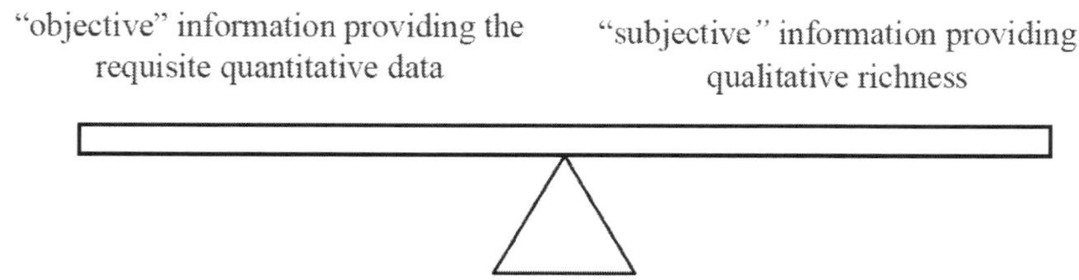

Figure 4. The Balance of "objective" and "subjective" information

Within the "In-Service" phase, numerous "subjective" and "objective" information sources are available for stakeholder exploitation, e.g.

- Subjective: Functional description, system inter- connectivity, modifications, concessions
- Objective: Operating performance criteria, e.g. temperature, pressure, mechanical load, electrical current, voltage,
- Stakeholders will often exploit information sourced from previous CADMID phases, e.g. Manufacture, for which information sources include:
- Subjective: Requirements specification
- Objective: Material used, dimensions, tolerances, test specifications.

Minimising "subjective" interpretation / analysis by exploiting the information available in a consistent manner should provide stakeholders with the most "objective" perspective of the information available.

4. "IN-SERVICE" STAKEHOLDERS

"A stakeholder is any entity (individual or organization) with a legitimate interest in the system" [7]; within the surface ship maritime defence domain the two key stakeholders are the RN i.e. "customer", and the COM, i.e. "service provider". Further analysis identifies a considerable number of stakeholders (Figure 5) with "an interest" in the vessel, i.e. operating, assessing, maintaining or validating its material state.

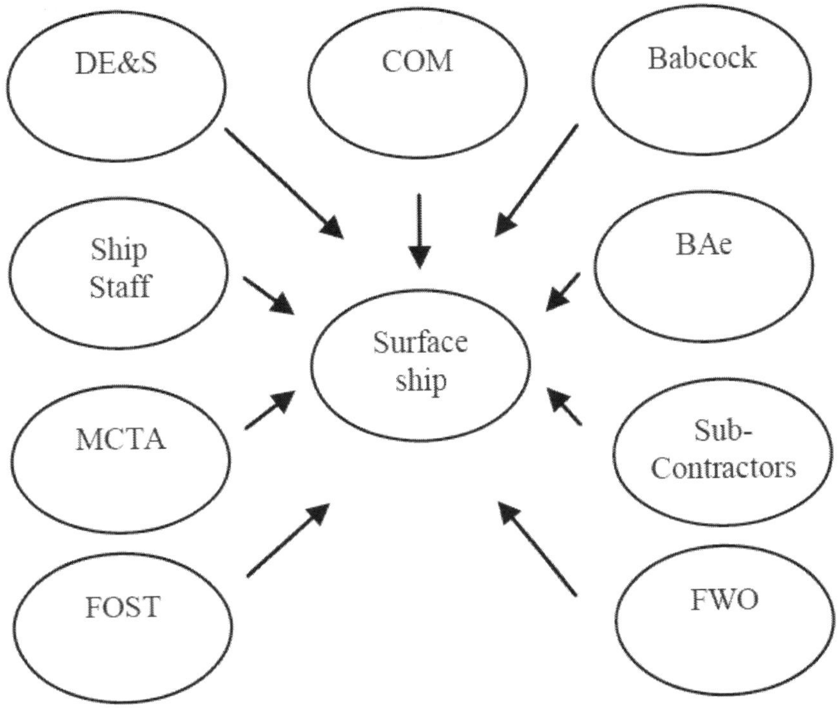

Figure 5. Key stakeholders involved during the "In-Service" phase of RN surface ships

A review of the "stakeholders' needs" reveals the following:

- Babcock: service provider and member of the SSSA, responsible for maintaining surface ships and submarines
- BAe: service provider and member of the SSSA, responsible for maintaining surface ships
- COM: organisation charged with maintaining surface ships, based upon the premise of "contracting for availability".
- DE&S: MOD's Defence Equipment and Support organisation, design authority and responsible for the purchase and support of all equipment and services of the RN
- FOST (RN): Flag Officer Sea Training, responsible for training and maintaining standards including whole-ship training ready for operations

- FWO (RN): Fleet Waterfront Organisation, support the
- vessel and assists in compiling work packages,
- MCTA (RN): Maritime, Capability, Trials and Assessment, responsible for performing dynamic trials and assessment of propulsion and auxiliary systems, ensuring engineering systems have met the requisite standard
- Ship Staff (RN): ships complement; responsible for
- maintaining and operating the vessel
- Subcontractors: numerous subcontractors providing services to Babcock and BAe in support of maintenance and assessment activities, e.g. Lloyds Register

The stakeholders' role of operating, assessing, maintaining or validating the material state exists within the cyclical mode of the vessel, i.e. Tasking, Upkeep and Regeneration (see below).

5. SURFACE SHIP "IN-SERVICE" MODES

As stated, the development and support of a warship will adhere to the CADMID lifecycle. During the "In-Service" phase a surface ship may be perceived as cycling through 3 discrete modes (Figure 6), i.e.

Note:

- Tasking: the period a vessel is under Naval Command HQ Scheduling Authority and thus available / undertaking operational tasking, referred to as "Fleet Time".
- Upkeep: a non "Fleet Time" vessel undertaking "deep"
- maintenance, typically including docking periods
- Regeneration: the timely activation, in full or in part, of existing force structures and infrastructure, including the restoration of manning, equipment and stocks to designated levels [17]

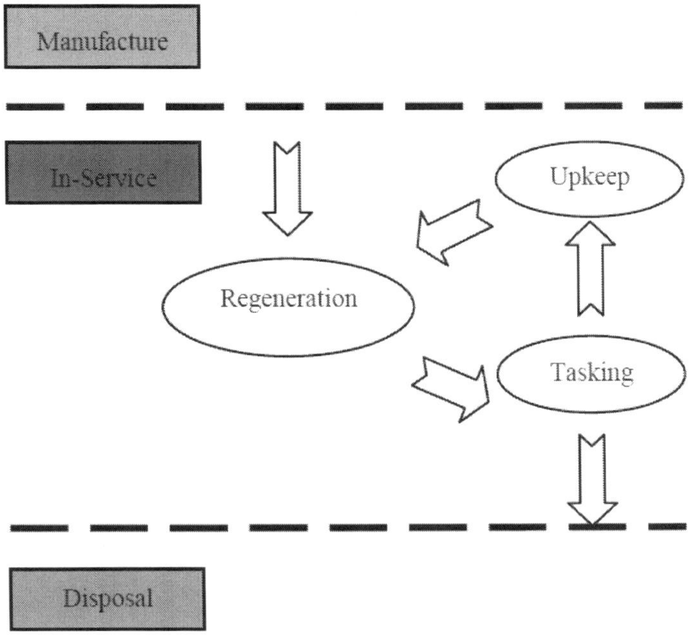

Figure 6. "In-Service" Cyclical Mode

The cycle of Tasking, Upkeep and Regeneration will be repeated throughout the life of the artefact; as intimated above, this may be significant, e.g. HMS Portland (Type 23 Frigate) ~ 34 years (planned).

It should also be noted the Tasking mode may be perceived as containing 3 discrete conditional "subsets" (Figure 7), i.e. Fight, Move and Float.

Each subset correlates to a functional decomposition of a vessels' capability, e.g.

- Fight: operation of offensive / defensive weapons and sensors
- Move: capable of self propulsion and maneuver
- Float: vessel able to remain afloat.

The information required, available and exploited within each mode clearly reflects the needs of the stakeholders and the demands of the vessel whilst in / out of "Fleet Time"; other factors may include the current / planned operational role, readiness state, shore support. In this regards, stakeholders will exploit information from each cyclical mode, however, the primary association for each stakeholder will typically be as shown in Table 2.

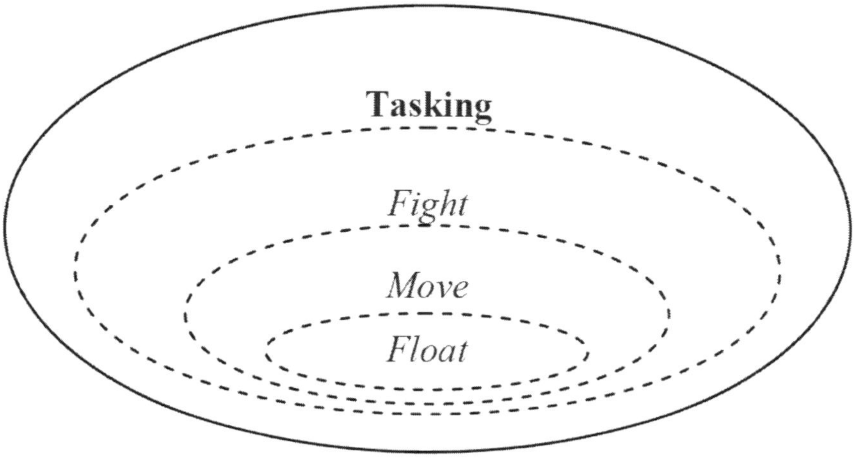

Figure 7. Tasking conditional "subsets"

Stakeholders apply and experience distinct "drivers" with respect to "the required characteristics and context of use of services" [10]. Whereby the "customer" is requiring / demanding "the most operational capable vessel possible"; the "service provider" is endeavouring to satisfy the customer's key requirement but within the constraints of a formal contract. The management of the interface between the "customer" and the "service provider" is undertaken by the COM, an organisation containing both "customer" and "service provider".

Table 2. "In-Service" Stakeholders and cyclical mode association

	Mode		
Stakeholder	**Tasking**	**Upkeep**	**Regeneration**
Babcock		•	•
BAe		•	•
COM		•	•
DE&S	•	•	•
FOST	•		•
FWO		•	•
MCTA	•		•
Ship Staff	•	•	•
Subcontractors		•	•

6. "IN-SERVICE" PHASE INFORMATION SOURCES

A considerable variety of information sources exist, however, the primacy of each (engineering) information source will vary with respect to the current cyclical mode, thus the key sources in each mode available for exploitation will encompass:

Tasking

- Engineering Logs: numerous logs detailing defects and operational information
- Health and Safety: information and procedures to ensure the safety of the vessel and ship staff
- Operational Defects: defects that impact the operational capability of the vessel
- S2022: defect reports

- Store Rob: information regarding requests for the removal of equipment from one vessel to satisfy a stores demand for a higher priority vessel

- UMMS: information related to concessions, preventive and corrective maintenance of systems; tasks may be categorised as, safety, environmental, operational or non- operational.

Note:

During Tasking, the information exploited and its associated characteristics will reflect the command aim, operational tempo and the vessel's situation within the Tasking "subset". For example, if the vessel's condition is such that it only exists within the subset "Float", the collation, analysis and exploitation of information will focus upon the immediacy of the vessels stability, trim and ultimately its capacity to remain afloat, e.g. HMS Endurance 16/December/2008 ~ hull valve failure, resulting in a flood of approximately $1000m^3$ per hour [18].

Upkeep

- Alterations & Additions: enhancements to vessel's capability, statutory modifications, e.g. navigation lights, alterations for operations / Fleet Time Fit, i.e. a vessel may be fitted "for, but not with" an equipment / system

- Contracts & Workspec: contracts and work specifications defining the work to be undertaken, including the use of "key" contractual words, e.g. Clean, Emergent Work, Inspect,... [19]

- Defects: defects necessitating dockyard support, material / stores, specialist test equipment, etc.

- Engineering Logs: numerous logs detailing defects and operational information

- HPUMA: Habitability Pre-Upkeep Material Assessment

- Material State Portfolio: report defining the material state on completion of upkeep, forms the baseline from which the next upkeep work spec. is prepared

- PUMA: Pre-Upkeep Material Assessment: assessment of the material state of the hull and systems, including the dynamic assessment of machinery, vibration analysis, thermographic survey.
- PUWSA: Pre-Upkeep Weapon System Assessment:
- assessment of the material state of weapon systems
- Set to Work: a complex series of trials, e.g. Harbour Acceptance Trial, Sea Acceptance Trial, Naval Weapon Harbour Test, Naval Weapon Sea Trial
- UMMS: information related to concessions and ongoing maintenance of systems, categorised as, safety, environmental, operational and non-operational.
- Project Reporting: standard project management reporting, e.g. Gantt charts, progress reports, financial reporting of expenditure, commitments, accrual,…

Regeneration

- Alterations & Additions: enhancements to system capability, statutory modifications,
- Engineering Logs: numerous logs detailing defects and operational information
- SARC: Safety And Readiness Checks: designed to assist the ship to manage the transition from being in dockyard hands to joining the operational fleet. This will include exhaustive Operational Sea Training, High Seas Firing, etc. such that the vessel will increase its "readiness state" and thus be "made ready to deploy from their home base or current location, appropriately manned, equipped, trained and supported".
- UMMS: information related to concessions, preventive and
- corrective maintenance of systems; categorised as, safety, environmental, operational and non-operational.

Note:

SARC is the primary task of the Regeneration serial, whereby the vessel is brought to full operational capability and readiness.

7. INFORMATION CHARACTERISTICS

The characteristics of the information detailed above will reflect not only the current cyclical mode, e.g. Upkeep, but also the urgency of the activity and whether the stakeholder is located onboard or ashore, e.g.

Onboard

- Technical: high technical content, reflecting the configuration and operation of the system
- Volume: high volume of data, necessitating maintainers and managers to "filter" noise
- Granularity: fine level of data granularity, e.g. logs detailing temperature / pressure / vibration…
- Command aim: direct and immediate, defining the aims and operational tempo, e.g. "Approached by two … the ships company realised how real the threat is." [20]
- Information availability: immediate: not only reflecting the command aim but also the proximity of systems and operational need
- Factual v analytical: primarily factual, with a degree of onboard analysis
- Subjective v objective: primarily objective, reporting the facts of an event

Ashore

- Technical: potentially reduced technical content, however, may be supported by the Original Equipment Manufacturer.
- Volume: potentially lower volume of data, filtered by ship staff and / or IT system
- Granularity: potentially reduced level of data granularity, often summarised.
- Command aim: removed from the immediate urgency of the operational task and threat

- Information availability: invariably delayed; in the case of ballistic nuclear submarines no contact is received from the submarine for up to 3 months
- Factual v analytical: primarily analytical, supported by factual information
- Subjective v objective: a mix of subjective and objective information, not only reporting the facts of an event but potentially an interpretation of that fact

Stakeholder information requirements with respect to the characteristics detailed may be further decomposed into:

Tasking Onboard & Ashore Upkeep Onboard & Ashore Regeneration Onboard & Ashore

Information may be exploited by merging and fusing the multifarious information sources, taking consideration of the information characteristics and balancing the subjective and / or objective nature of the content.

"In-Service" information in each of the modes may be perceived as a jigsaw image; whereby the individual sources correspond to the tiles of a jigsaw (Figure 8), unfortunately, some tiles may be missing or even damaged, none-the-less the stakeholder must exploit what is available to provide the optimum "image" of the vessel, system or plant item. Charles Handy reinforces the perception of a "jigsaw", when he writes:

"Information, above all else in life, seems to display the essential features of synergy. The whole is so often more meaningful than the sum of individual parts. An information jig-saw, even though all the pieces are separately available, is nothing until put together." [21]

Note: MIMIC - Vibration monitoring system

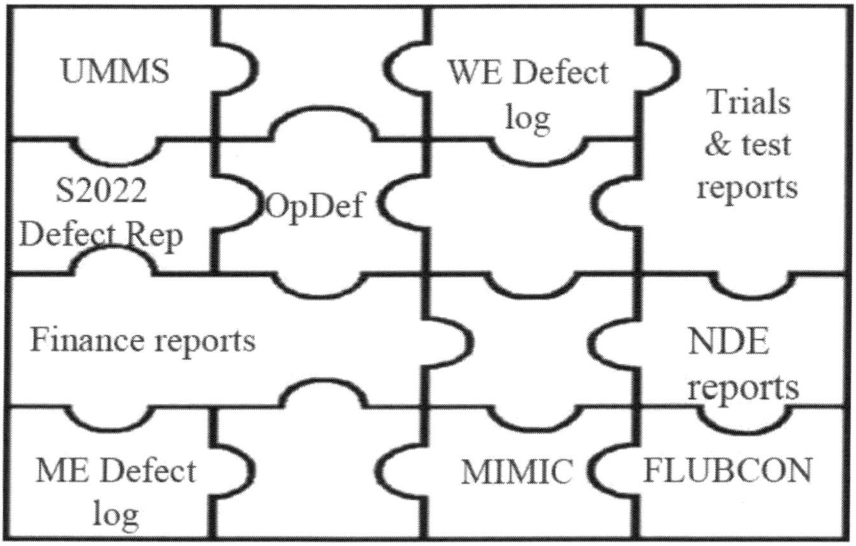

Figure 8. The jigsaw of "In-Service" information sources

Tacit knowledge will assist in interpreting / identifying the jigsaw "image" and will potentially compensate where confidence is low. The value of tacit knowledge should not be overlooked; since 90% of the knowledge in an organisation is considered to be in peoples' heads [22, 23, 24].

8. DISCUSSION AND FURTHER RESEARCH

The volume of information available should not be under- estimated; thus a significant area of research is identifying the key data elements within each cyclical mode that are relevant to the stakeholder and indicative of a vessels' / system material state. The scope of the research will include the availability and readiness of the vessel to undertake operational tasks.

Variances in the characteristics between onboard and ashore information sources may inhibit the response and capability of stakeholders. However, the research will not only highlight the key data elements but their relevance to the stakeholder being onboard or ashore.

Tacit knowledge provides stakeholders with the intellectual capacity to exploit multiple information sources for complex systems and disregard information when and where appropriate, e.g. data anomalies within electronic systems.

The logic of exploiting information is invariably experiential: typically, stakeholders will formulate an assessment applying Procedural knowledge (as contained in rules and regulations) and Declarative knowledge (as formed by problem solving) based upon subjective and objective information sources.

Hence, a further area of research would be the influence of knowledge sources upon information exploitation (Figure 9),

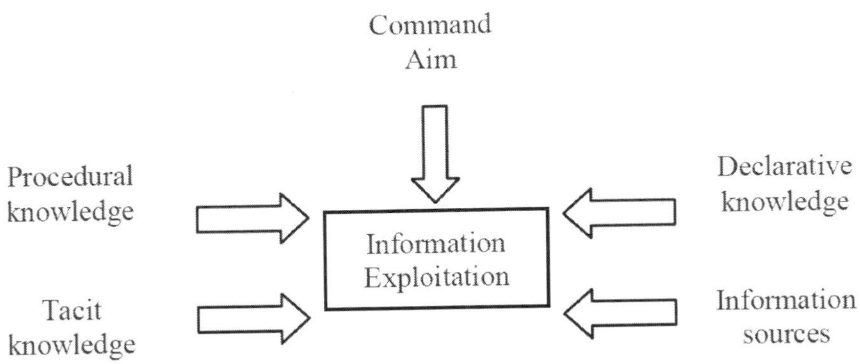

Figure 9. Information exploitation influences

9. CONCLUSION

The "In-Service" phase of a surface ship includes both operation and maintenance; these activities typically cost an organisation most of the entire lifecycle costs. In some cases up to 70% of the total cost of ownership of an asset is consumed at this stage [25].

The exploitation of information within the "In-Service" phase may be considered to be a "wicked problem", e.g.

- Changing "command aim"; resulting in a switch in
- operational focus and tempo
- Complex system of systems
- Subjective / objective perspective of information
- Cyclical mode change, i.e. Tasking, Upkeep and Regeneration
- Numerous information sources

Currently each of the stakeholders are dependent upon Suitable Qualified and Experienced Personnel assessing and exploiting information, optimising their flexibility and ability to learn, however, this is a limited and expensive resource.

The research indicated above will assist in identifying the key data elements which will enable the stakeholders to prioritise the collation and exploitation of information within the marine surface ship domain.

ACKNOWLEDGEMENTS

The author is grateful for the support of the EPSRC-funded Bristol/Bath Industrial Doctorate Centre in Systems.

REFERENCES

1. SIA, 2013. The Collins Class. http://www.submarineinstitute.com/submarines-in-australia/The-Collins- Class.html [Accessed: 20/Jan/2013]
2. Kane, G.C., Alavi, M., 2007. Information Technology and Organizational Learning: An Investigation of Exploration and Exploitation Processes. Organization Science; Sep/Oct 2007, Vol. 18 Issue 5, p796-812
3. Blockley, D., 2010. The importance of being process, Civil Engineering and Environmental Systems, 27:3, 189-199

4. UK Defence Statistics 2012 – Formations and Vessels http://www.dasa.mod.uk/modintranet/UKDS/UKDS2012/c4/table40 1.ph p [Accessed: 20/Jan/2013]

5. New, C., Steven, G., 2011. The contradictions in modern naval asset management and their resolution. In *Asset Management Conference 2011, IET and IAM* (pp. 1-5). IET.

6. Babcock, 2012. Transforming Warship Support. http://www. babcockinternational.com/media-centre/transforming- warship-support/ [Accessed: 20/Mar.2013]

7. INCOSE, 2010. Systems Engineering Handbook. A Guide for System Life Cycle Processes and Activities. Ver. 3.2 INCOSE, San Diego, CA. January/2010.

8. Honour, E.C., Valerdi, R., Initiative, M.L.A., 2006. Advancing an ontology for systems engineering to allow consistent measurement. In *Conference on Systems Engineering Research.*

9. MOD-AOF, 2013. Acquisition Operating Framework. https:// www.aof.mod.uk/index.htm [Accessed: 12/Mar/2013]

10. ISO15288, 2008. ISO/IEC 15288:2008 Systems and software engineering -- System life cycle processes. Organisation for Standardisation, Geneva, 2008.

11. MOD, 2013. HMS Somerset returns to sea after £20 million refit. https://www.gov.uk/government/news/hms-somerset-returns-to-sea-after-20-million-refit [Accessed: 27/Jan/2013]

12. MOD, 2009. The Systems Engineering Handbook "Principles, Practices and Techniques"

13. OGC, 2005. Common Causes of Project Failure. UK Office of Government Commerce, HM Treasury.

14. Soderholm, P., Holmgren, M., Klefsjo, B., 2007. A process view of maintenance and its stakeholders. Journal of Quality in Maintenance Engineering Vol. 13 No. 1, 2007 pp. 19-32

15. Giurgiutiu, V., 2007. Damage Assessment of Structures-An Air Force Office of Scientific Research Structural Mechanics Perspective (Vol. 347, pp. 69-74).

16. Morgan, G., Smircich, L., 1980. The Case for Qualitative Research. The Academy of Management Review , Vol. 5, No. 4 (Oct., 1980), pp. 491-500

17. AAP-06, 2013. NATO Glossary of Terms and Definitions (English and French). 03/April/2013.

18. Endurance, 2010. Service Inquiry into the flooding of HMS Endurance. https://www.gov.uk/government/publications/service-inquiry-into-the- flooding-of-hms-endurance [Accessed: 20/May/2013]

19. BR8593, 2012. Procedures for Compiling and Processing Work Packages for Upkeep Periods in Surface Ships. Vol.12. MoD (Navy),Abbey Wood. July/2012

20. HMS_X, 2010. HMS X, Ships Log S322. March/2010

21. Handy, C., 1999. Understanding Organisations. 4th edition 1999,Section 2.4, pp129

22. Wah, L., 1999. Making Knowledge Stick. Management Review. May pp24-9

23. Bonner, D., 2000. Knowledge: From Theory To Practice To Golden Opportunity. American Society for Training & Development.September-October. pp12-13

24. Lee, J., 2000. Knowledge Management: The Intellectual Revolution. IIE Solutions. October, pp 34-7.

25. Koronios, A., Nastasie, D., Chanana, V., Haider, A. 2007. Integration Through Standards - An Overview of International Standards for Engineering Asset Management. 2nd World Congress on Engineering Asset Management and the Fourth International Conference on Condition Monitoring (WCEAM 2007). Harrogate, United Kingdom.

Index

H

I

J

M

N

O

P

R

S

T

U